The Rationality of the Christian Faith and the Rationality of Science

Paul Peter Rom Abim

The Rationality of the Christian Faith and the Rationality of Science

Understanding Stanley Jaki

PETER LANG

Bibliographic Information published by the Deutsche Nationalbibliothek
The Deutsche Nationalbibliothek lists this publication in the Deutsche Nationalbibliografie; detailed bibliographic data is available online at http://dnb.d-nb.de.

Library of Congress Cataloging-in-Publication Data
A CIP catalog record for this book has been applied for at the Library of Congress.

ISBN 978-3-631-87611-4 (Print)
E-ISBN 978-3-631-87612-1 (E-PDF)
E-ISBN 978-3-631-87613-8 (EPUB)
DOI 10.3726/b19610

Peter Lang – Berlin · Bern · Bruxelles · New York · Oxford · Warszawa · Wien

This publication has been peer reviewed.

www.peterlang.com

To all those who seek to understand more deeply
their Catholic Christian faith through science and philosophy

Acknowledgements

The Acoli people say *'ngininingini coka'*, which contextually means even the tiniest insects, can do a great job, when they contribute their efforts to a desired goal. I must say that this book would not have been a success without the contributions of many people. I will only acknowledge those whom my memory will allow me to, with an assurance to those who will not see their names that they are deeply acknowledged even by the fact of noticing that they are missing on the list.

My first gratitude goes to God for all the protection that he has accorded me from the day I was born, through the terrible moments of war in Northern Uganda, through the Ebola time in the seminary, the tough times of the coronavirus pandemic from 2020 to date and throughout my entire life. I thank my parents who have shared with me the life God gave them and being teachers taught me to work hard and never to settle for less than my best. Above all, I thank them for passing unto me the Catholic Faith. Mum and Dad, you made me exploit my potential to the limits. I thank all my siblings with whom I share life: Reverend Mother Angelina Grace Aciro, LSMIG, Easter, Severino (RIP), Paska, Baptist, Lucy, Michael, Francis, Irene, Teddy, Francesca, Benjamin and Jacinta.

My gratitude to my Bishop Most Reverend John Baptist Odama, who trusted me and sent me back to Rome to finish my formation in Philosophy. I thank the Congregation for the Propagation of Faith for giving me scholarships on two occasions for my studies.

In a very special way, I would like to thank with my whole heart Professor Giambattista Formica, who supervised my Doctoral dissertation with patience and gave me the impetus to dig deep into philosophy. I thank Professor Lorella Congiunti, who has been my inspiration way back in 2010. Professor Congiunti, you have always given me euphoric moments from rigorous intellectual work. Every time I open my eyes and behold nature, I am filled with awe and stupefaction because you taught me to understand nature in the philosophy of nature both intensively and extensively.

I owe gratitude to Antonio Colombo, for his promptness in providing the resources I needed about Stanley Jaki, for inviting me for conferences and also enrolling me as part of "Stanley Brigade." You kept me updated with every new publication about Jaki. I thank Fr. Rafael Pascual, for exposing me to the resources of Jaki in the Pontifical University Regina Apostolorum and for the initial indications of electronic resources of Jaki.

My gratitude to Massimo, Paola, Cinzia, Francesco and Maria Chiara, for the support they have accorded me since my seminary days. In the same wake, I thank my American friends, beginning with Mike and Sue Wood, Bobbi Vaughn, Bill, Todd and family, Mark and Anita Whims, Matt and Chris Dunbar, and Hope for Uganda for prayers and financial support towards the cost of books, flights, access to American libraries and living in Italy. Special thanks to *Nini* for being my priceless American secretary struggling to read my poor calligraphy. Gratitude to Msgr. Matthew Odong who has always been a source of encouragement for all my academic endeavours and also to my German family, Mamma Klara, Reinhard and Birgit. Gratitude to my English friends, especially Rev. Fr. Simon Peter Firth, Marcia, Marion, Brigit Cooper and all the parishioners of the Parish of St. Helen and the Holy Cross in Ashton, the diocese of Salford.

I thank the Rector, Staff and students of Alokolum National Major Seminary, Sacred Heart Minor Seminary Lacor and Archbishop Flynn Secondary School, for providing me the necessary tranquillity for pursuing deep academic work. Thank you, Rev. Fr. Dr. Augustine Mugarura, for being a light to my academic path. I thank my former Rector Rev. Fr. Dr. Paul Masolo who whispered into my ears at the end of my theological studies: "Love the Church, the sky is the limit." These words remained with me all the time especially when things got tough. Gratitude to Rev. Fr. Dr. Richard Buyinza and Rev. John Felix Opio, for proof reading my work despite their heavy respective schedules. Thanks to the Parish of St. Mary's in Ambra: Don Pier Luigi Cetoloni, Piero and Rita, Marco Rossi and family, Alessandro Artini and family, Lorenzo Guarnera and Patrizia, Roberto Fantoni, Marinella and all the parishioners for the life and the word of God shared for the six years I have been in Italy, at two different occasions. Thank you to all those who have contributed in one way or another and are not mentioned here. You know that I am grateful for your contributions, even when my memory cannot allow me to mention you here by names.

Table of Contents

Abstract

To deal with rationality is to immediately deal with reason as the fundamental distinguishing mark of the human person. Reason and its rationality make us deal no longer with a thing but a person, whose existence precedes his thinking and whose thinking be it religious or scientific is according to that very faculty of reason. It is therefore, justifiable to deal with science and religion as two realities distinct but integrated by human reason. Science and Christian faith practiced in religion are two human occupations that have gained central attention of the core of the existence of the human being himself. In science, man tries to decipher the content of nature, what it is made of, how it functions and how he can use them for his benefits and the benefit of others. Scientific investigations, therefore, become a natural mission to conquer and subdue the world and render to God the service of the totality of creation. The investigation into natural things by science is only possible because man is reasonable and, because he is reasonable, he finds some patterns in nature that demonstrate it being a product of reason. It is this reasonableness that constitutes the rationality of his operation. Man's other natural ability is the transcendence of the natural things that object to his intellect and senses. This transcendence leads him into metaphysical questions, the most fundamental of which is why there should be something at all and not nothing. He relates concretely to this reality beyond him in the practice of his religion. Man's religious attitudes seems to be a natural inclination to look at realities beyond the physical. It is this metaphysical desire that drives him to seek the cause of his own being and that of all beings. These metaphysical questions cannot be arrived at without the use of reason. Nature is so rational that man cannot but see beyond it in his rationality. Religion and science therefore, have two common denominators: reason and natural things. If this is so, and if reason is only one, then there must be an integration in the one without fusing the two. If this is possible, then philosophy, that is the love of wisdom cannot but be the unifying master that will not be ignored. These words of Etienne Gilson summarised it all:

> Where it in my power to do so, I would rather leave you with a gift. Not wisdom, which I have not and no man can give, but the next best thing: the love of wisdom, for which philosophy is another word. For to love wisdom is also to love science, and prudence; it is to seek peace in the inner accord of each mind with itself and in the mutual accord of all minds.[1]

Stanley Jaki delved deeply into philosophy from his study of the history of science after reading the beautiful book of Gilson, *The Unity of Philosophical Experience* (1937), a book that demonstrate one of the best of arts of doing philosophy. It is wisdom and prudence that guarantees the equilibrium both in doing science and in the search of the ultimate explanation of all things in religion and the love of that wisdom plays a bridging role between science and religion.

1 E. GILSON, *The Unity of Philosophical Experience,* Ignatius Press, San Francisco (CA) 1937, 257.

General Introduction

Many people, even highly educated academics, when they hear of someone studying philosophy, retort almost instinctively: "will you have your feet on the ground?" What will studying philosophy offer to the world? What work will you do after studying philosophy apart from teaching? These are questions that are a clear demonstration of the misunderstanding of what philosophy is and its mission. The mission of philosophy is to seek and consequently communicate the truth in a more systematic, universal and fundamental way. Being the love of wisdom, we acknowledge that wisdom does not come easy. In a world that no longer understands the difference between right and wrong, just and unjust, truth and falsehood, science and non-science – a world in which relativism reigns even for highly learned people, the need for philosophers is arguably greater than ever before. The true vocation of philosophy since time immemorial has been the provision of equilibrium using a rational discourse.

The issues of science and religion have lingered in my mind for over 15 years now, especially when I see the confusion and misconception that continue to pass from generation to generation about science making religion irrelevant. I noticed also that the greatest force behind the secular world is the appeal to science as the alternative explanation of all things, as opposed to the traditionally held view of religious explanation. In the words of Jaki, "science was set up [by enlightenment] as the great debunker of the supernatural, and a proof that anyone engaged in that debunking almost automatically counted as an expert of science."[2] This ideology drives the secular world to an either-or mentality, that is, either religion; the sole authority that can explain why we are here, and what we deal with in nature, or it is science who has the final word. To have no middle ground or point of intersection is reductive and dangerous in the search for a holistic explanation of reality. The biggest question is whether people who think so have bothered to understand science and religion, what they stand for

2 S. L. Jaki, *A Mind's Matter: An Intellectual Autobiography*, Grand Rapids (MI) 2002, 43.

and their origins in order to have a fair judgement of their relationship and concerns. In order to be a judge, one must thoroughly inform oneself of all matters concerning a particular case. There is a lot of information about the relationship between science and religion and yet there seems to be growing problems regarding the same. What is the problem? How can we understand the relationship between the two without having to belittle or to eliminate any of them? Could philosophy take up the role of bridging the two streams? Would the elimination of one for the other not directly militate against reason itself?

I soon realized that there is a common denominator that is either ignored or poorly employed in the search for the truth. My principal aim is to argue that science and religion cannot be in conflict, as long as they are both rational human efforts in search of meaning in nature and life. The source of the apparent conflict between science and religion would then not be in science and religion themselves, but in the mode in which they are studied and propagated; bluntly put in modern terminology, it is 'user problem.' I would like to demonstrate that reason, being the distinguishing mark of man, is the converging point for a fruitful dialogue between Christian faith and modern science. To take a conflicting position or an exclusive position is to defeat the purpose of reason itself and alienate it from its proper function of unity. If faith is reasonable and science uses reason to investigate into quantitative reality, then harmonic dialogue should be a great option in the relationship between modern science and Christian faith. I specifically used "Christian faith" because of its intellectual rigour as opposed to superstitions, justified by Jaki thus: "of all faiths, Christianity alone has specific formulas known as creedal statements."[3] This is to stay that Christianity has an intellectual formulation of what can be taken as deposits of faith found throughout Church documents and salvation history. I would not like to fall into the traditional arrogance that Christianity is a superior religion or that other religious believes are not rational. My choice of Christianity is motivated by being a Christian myself and also as already stated because the Christian faith is rationally defined in its dogmas and creed. It is convincing to me that the Catholic Church, to be specific,

3 *Ibidem.*, p. 125.

propagates faith without contradicting reason. This makes the Christian faith a rational enterprise. When Christians argue from ignorance or cannot defend their faith rationally, it is the failure of Christians and not faith being irrational.

I was greatly influenced by the works of Richard J. Connell, *From Observables to Unobservable in Science and Philosophy* (2002). This work was introduced to me by the daughter of the author Anita Whims, who claimed to have never understood much of what her philosopher father wrote. A very humble way of acknowledging the great works of a staunch Catholic professor of philosophy. The affirmation of Connell on the fact that observables lead us to the unobservable shows clearly the power of the mind to transcend itself, alluding to the fact that realities do not exist only because we can observe them, making observables only a small part of reality that science deals with and in a constantly provisional way. Later I got exposed to the following books by the same author: *Matter and Becoming* (1966), *Logical Analysis: An Introduction to Systematic Learning* (1981), *Substance and Modern Science* (1988), and *The Empirical Intelligence – The Human Empirical Mode,* (1988).

The central argument I will be dealing with in this exposition is the rationality of science and the rationality of Christian religion. I chose the works of Stanley Jaki because of its vastness and yet its limitation to one concrete question, the historical reconstruction of the relationship between science and religion and the limits of science. This was attractive because being a scientist and a theologian, Jaki would be the best candidate to defend what he believes and practices. Another motivation was the fact that Jaki historically proved that science, right from its birth, had a lot to do with Christianity. This was a bold stand to take in a secularised world, where politically correct language does not spare even academics. This is important for me because it points to the fact that the apparent conflict must have a better explanation, if a good understanding of the relationship between science and religion is to be reached.

One of the responsibilities of a philosopher is to define his terms to avoid confusions and misunderstanding. I would like to clarify some terms that should be taken univocally throughout this book. The term rationality carries different meanings in analytic philosophy. This is the reason why I would like to define what we mean by rationality in this book. In this exposition, rationality refers to reasonableness, the driving principle, the ability to use reason well or rationale behind science and Christian religion. The term rationality is also taken to mean the capacity to reason and think in an organised way that forms within itself an internal coherent structure. It is therefore not enough to have the rational and thinking capacities but also a necessity to think and reason correctly. For reason not to be contradictory, it must adhere to the rigour required of it; that is, it must seek the truth, present it wisely and logically. Christian faith therefore is rational in the sense that what is believed is true and can be rationally presented for the comprehension of a third party. Rationality in this book should not be taken at any rate to mean rationalism which is a form of reductionism. Rationality should be taken to mean an honest use of reason to understand rational things by science and to transcend from the same nature to the realm of religion. Science and religion both need reason to acquire the knowledge of their contents and therefore, according to me, the examination of the unity and operation of reason is the best way of establishing the relationship between science and religion. As reason is gift to man by his very nature, his rationality should be cultivated because it is the mode in which the human being uses his reason correctly to separate and unite different things into a coherent body of knowledge. When we talk of faith, we shall be referring explicitly to the Christian faith, which is a belief in a personal and responsive God revealed in Jesus Christ, unless it is qualified or otherwise indicated to mean something else. When I talk about some form of faith in reference to science, it should not be taken in the same way as the faith of a Christian coming from a revelation but a strong belief that conditions necessary for science will always remain constant. It can also mean trust in the consistency of the law of nature. It is true that the term faith can carry many meanings and therefore the context of its use in this book will determine the sense to be applied to it, apart from faith as a revelation. Science, on the other hand, will refer to that for which a person can be called a scientist and the interplay between human intelligence and reason in experiments, that is what Jaki defines as 'exact science.' This makes the human being both scientist and a person of faith. Sometimes science should be understood as the rigorous, methodical and systematic study that covers other disciplines that fall outside experimental sciences. The context in which the term science is used will determine what it should mean.

I set out in this book to reconstruct philosophically the arguments for the harmonic and integral relationship between science and religion in the footsteps of Stanley Jaki. It is my intention to argue on the role played by philosophy as a bridge between science and religion. Being rational, scientist cannot therefore escape the requirements for acting according to the dictate of reason and therefore responding to ethical demands in their investigations. Jaki, however, sometimes expresses his fear that philosophy has lost its very principles of the love of wisdom. "Had philosophy not ceased to be a pursuit of the love of wisdom, it would perhaps have been able to serve as a saviour for that much needed catholicity."[4] As much as this is an invitation to the practice of good philosophy, it is arguable whether philosophy as the love of wisdom has metamorphosed. It can be contented that, as long as those who call themselves philosophers no longer seek wisdom, they cease to be philosophers and consequently what they do can be best described as ideological dissemination. In short, philosophers can falter in their thinking and base their philosophy on flawed principles while solid philosophical principles remain intact. Philosophical mistakes do not constitute the failure of philosophy but philosophers. In his earlier work, Jaki had hinted to the fact that fake philosophers can advocate for dubious philosophy.

Jaki himself provides an antidote to this fall by saying: "genuine respect for man's mind, a respect without which his sense of purpose will be at best anaemic, should be sought in perspective of the mind in which its need for matter is fully recognised."[5] Extreme rationalism or empiricism contradict rationality for which balance and harmony is intrinsic. Man is neither exclusively his mind nor exclusively matter and such a fall is not the fall of philosophy but philosophers. It is my desire to reaffirm the need to continue to offer good practice of philosophy that is reasonable to promote fruitful and harmonic dialogue between science and religion.

---

4 Id., *The Saviour of Science,* William B. Eerdmans, Grand Rapids (MI) 2000, 171.

5 *Ibidem.,* p. 172.

The enduring character of truth is so powerful that every researcher at one point should make a choice between being subjective or sticking to the objective truth dictated by reason. It would not be out of place to state right from the onset that such a choice is what I have made when I set out to carry out research on the religion-science relationships. We shall allow ourselves to be guided and directed by objective reason that defines and separates the human person from all other beings.

Over the centuries, the history of science has taken different dimensions depending on the interest of the historian. Some scholars took science as a systematic rigorous discipline as their line of thought and therefore talk of sciences and not just science, while others want to merely follow the line of physics or positive science as true science, and therefore talk of science in the singular form.

The error of every age and time in the world of thoughts and history has always been the lack of balance and reductive approach to knowledge. Jaki himself tackles the problem of reductionism in the history of science. Reductionism in philosophy is the limitation of knowledge exclusively to a particular field and method. Jaki argues that "the modern times praise and worship reductionism as long as they seek to eliminate any metaphysical discourse in their studies."[6] It is actually fashionable to be anti-religious or anti-metaphysics in our modern world. Sam Harris, Richard Dawkins and Christopher Hitchens are great testimonies of our time and Jaki, aware of the challenge of our modern times, writes:

> In the intellectual climate of our times respectability is reserved to such avenues of reason as are meticulously free not only of theological connotations but of even the least touch of metaphysics. The novelty lies in the resolve to restrict rational discourses to the pattern set by methods of physics.[7]

It is hugely controversial problem that we shall be dealing with, when we shall discuss the problems of method and intellectual honesty. I will use the example of the above modern atheists to demonstrate the dire need to do a deep study on the relationship between faith and science. These modern

---

6 Id., *The Roads to Science and the Ways to God*, Real View Books, Port Huron (MI) 2005, 214.

7 *Ibidem.*

atheists use science in very compelling but shallow way to belittle religion and faith in God.

Science and religion have gone and continue to go through these extremes even in our modern times. More often than not, the two are put into confrontations that are uncalled for by those who play the politics of science and religion. Many true scientists seem indifferent or at least contented with their sciences but many take science as an alternative of faith, and thus are creating problems and confusions that are technically uncalled for. Time and history have proved that religion and science can coexist without conflict, as long as they both respect their limits and remain intellectual enterprises carried out by rational and relational beings.

The relationship between science and faith is a complex one that demands a deliberate employment of a rational rigor as the first requirement. It demands a thorough philosophical formation to fully comprehend the complex relationship between the two. Even when in the American society the first reaction, generally, seems to be that there is no conflict between faith and science. The reason is simply clear that many intellectuals do not take matters of faith seriously anymore and the underlying reason is that science seems to answer most of the questions related to physical problems of day-to-day life and faith in God seems to set ethical standards so high for a pleasurable life. There is still in the modern times the Vienna Circle mentality that whatever cannot be proven empirically is childish and profession of ignorance or what Jaki calls the unrealistic claim of some scientists that "science is the only valid underlying knowledge that gives guidance to the adventure."[8] These extremes have the risk of always creating problems and conflicts because they are intolerant to other views.

In fact, the anti-metaphysic's mentality gave birth to scientism. Scientism was greatly promoted by the Vienna Circle as a form of reductionism that held that only positive science with its methods is the true form of knowledge. Scientism is a deteriorated form of science that lacks both philosophy and science itself. It is an ideology against metaphysics that ended up with time a ridicule to science and its methods. For Jaki, "scientism is not a genuine reverence for science but harnessing of science for what science is

8 Id., *The Saviour of Science*, 206.

not."[9] In promoting science to do what it is not and downgrading all other forms of knowledge, conflicts become inevitable.

Edward J. Larson and Michael Ruse, in *Faith and Science* (2017), argue that the term conflict is used analogically and that "the recognition that its application to the relationship between science and religion is both seductive and tenacious suggests that many people respond to it and see it as valid."[10] We cannot therefore ignore the demand for a thorough and clear investigation into the fruitful relationship between faith and science.

Division of the Work

This book is divided into three chapters of relatively equal length. Chapter One deals with the arguments for the viable birth of science in the later medieval times and in the Christian civilisation in the West. In it, we looked at the failure of science in some ancient civilisations, such as Egypt, Mesopotamia, Greece, China, India, and among the Muslims despite great technologies and scientific discoveries in some of them. The argument here is that science needed to be "fanned into a sustainable flame" for a viable birth and that a flash of "scientific light" was not enough to constitute a scientific revolution that we now have. Science needed to break away from the ancient world view that enclaved it in eternal return in the name of the 'great year' and the pantheistic vision that impeded scientific innovations, in order for it to achieve a viable birth worthy of its current state. It was only Christianity, according to Jaki, with a clear separation between creation and creator that could viably give birth to science. Christianity had the advantage of setting up the mind-set and providing the institution and education that promoted science. That Christianity played a great role in the birth and development of science is incontestable, but to assert that without Christianity science could not have been viably born, remains a controversial issue. Anyone venturing into this discussion ought to understand what Jaki means by a viable birth.

---

9 Id., *The Road to Science and the Ways to God*, 218.

10 E. J. Larson – M. Ruse, *On Faith and Science*, Yale University Press, New Haven (CT) 2017, 6.

The second chapter deals with the thorny problem of the relationship between science and religion, which directly affects faith in God as the basic component of religion. It tries to explore this relationship as traditionally and historically classified by Ian Barbour in order to situate the arguments of Jaki in the various models. Collocating Jaki's thoughts about the relationship between science and religion in Ian Barbour's models goes a long way in achieving our objective of philosophically reconstructing Jaki's thoughts. Jaki proposes three major things for a good relationship between science and religion. First, that the two must be correctly and strictly defined and in doing so setting boundaries and limits that should be respected. Secondly, that the two are conceptually distinct and no attempt whatsoever should be made to mix them. What God has put in conceptually different domains; no man should mix them up. Fusing them up has been responsible for a legion of confusions. Thirdly, that philosophy is an indispensable requirement for the proper understanding of the relationship between science and religion. The two streams can only be hovered over by reason and bridged by philosophy because it [philosophy] has the method necessary for it.

The third chapter deals with the unity of rationality. It is an invitation to obey the requirements of reason that calls for ethical concerns on the part of the scientists not as stemming from religious convictions but from the dictates of reason itself. It is an invitation to the rigorous use of reason in the world in which it seems optional. It also treats some misconceptions that arise about religion that may not necessarily concern it in the strictest sense. It is my conviction that when reason is put into good use, all rational beings capable of reason can transcend from nature into the metaphysical realm, without losing touch with concrete reality. Chapter Three leads us into some general observations, recommendations and conclusion. My central argument is that if man is rational and capable of rationally investigating into nature, and if the same man has the eyes to see that the same nature leads him to spiritual realities expressed and practiced in his religion, then there is a possibility of a harmonic relationship between science and religion and an intellectual integration in the rational power that belongs to human beings alone.

A Brief Biography of Stanley Jaki

Owing to the fact that the biography of Jaki does not form part of this research, for a simple reason that it is easily available electronically, I would, like to briefly give some information about our author.[11] First of all, Stanley Jaki was a priest who was a scientist, theologian, historian, and a philosopher born in August 17, 1924, in Győr, Hungary. He obtained his PhD in theology way back in November 1950 from Saint Anselm University in Rome and another doctorate in physics in June 1957. It was at this time that he became interested in the history of science, as a means to settle or rather comprehend well the issues between science and religion. His first major publication in the line of the history of science was *The Relevance of Physics* (1966) that became the springboard for many of his subsequent publications. The publications relevant to this exposition are found in the list of bibliography. Jaki loved to research and write. This was partly conditioned by his health when he lost his voice and could not continue teaching. The best way he could communicate without his voice was through his writings. He was greatly influenced by Pier Duhem on whom he researched and wrote extensively. This was in the wake to trace the medieval roots of modern science. He was influenced too, in his intellectual journey by Gilbert Keith Chesterton, Etienne Gilson, Jack Maritain and John Henry Newman among others. Jaki had contacts with prominent scientists and scholars of his time, like Karl Popper, Paul Winger, Robert A. Nisbet, Eugene Wigner, Jerome Lejeune, to mention a few leading authorities and various Nobel prize laureates. These personal contacts shaped a lot of his thoughts on the subject of science and religion, as he met both those who challenged his

---

11 For a detailed biography of Stanley Jaki, cf. P. HAFFNER, *Creation and Scientific Creativity: A Study in the Thoughts of S. L Jaki*, Gracewing, Leominster, 2009, 11–23. For details and motivations about his publications, cf. S. L. JAKI, *Minds Matter: An Intellectual Autobiography*, William B. Eerdmans Publishing Company, Grand Rapids (MI) 2002. For details of his character and relationship towards the end of his life, cf. R. PASCUAL – A. COLOMBO – J. BEAUMONT, "Who was Fr. Jaki?", in M. GUERRA – R. PASCUAL – A. COLOMBO (eds.), *Science and Faith in Stanley Jaki. Proceedings of the Summer Course*, If Press, Rome 2017, 13–34.

thoughts and those who praised and gave him professional and specialised insights. He was continually in search of knowledge from primary sources.

He was appointed as an honorary member of the Pontifical Academy in 1992 by Pope John Paul II. He lectured summer courses on science and faith at Pontifical University Regina Apostolorum, Rome until the time of his death. He died of a heart attack on 7 April 2009, in the intensive care unit of the *Clinica de la Conception* in Madrid, Spain. He was on his way back to the USA after delivering a lecture in Rome. Stanley Jaki can be described as a resolute Catholic priest who sought to defend the Catholic faith using science, philosophy, and theology. He is one who lived the integration of science and religion in one man. He is stern and sometimes polemical in his vast writings that demonstrates wide range of knowledge.

Chapter One The Concept of Science in the Context of Western Civilisation and the Debate between Science and Faith

1.1 Introduction

The fundamental question I would like to tackle here is the root of science in Western Civilisation. This Western Civilisation at one point was the same as Christian civilisation. Stanley Jaki, historian of science, sustains a strong argument for the singular viable birth of science in the Christian West in the late medieval times that has eventually constituted what is known as Western Civilisation. I am particularly interested in the philosophical arguments as to why this was not possible in other civilisations despite the fact that they had great potentials and opportunities.

Many theological arguments have been raised on the thinking of Stanley Jaki on the relationship between faith and science or religion and science. Paul Haffner has sought to put to light some theological thoughts of Jaki relating it to the teachings of the Catholic Church and some contemporary theological issues. He acknowledges the need for a solid philosophy in handling matters concerning the relationship between science and religion. In his own words he says:

> The dialogue between faith and science has a foundation, on ontological ground, in the fundamental relation between science and creation. A right understanding and good vision of faith in God the Creator with a solid philosophical and cultural grounding could lead to a more effective dialogue by the Church in the modern world with scientists.[12]

This is pertinent for us because we shall use philosophical arguments to try to understand issues concerning Christian faith, science and the principles that govern human intellect in relation to the created world. Needless to say, the philosopher has a role to play in unifying the various bodies

---

12 P. Haffner, *Creation and Scientific Creativity: A Study in the Thoughts of S. L Jaki,* Gracewing, Leominster, 2009, 6.

of knowledge to some coherent form that satisfies good reason. Haffner's *Creation and Scientific Creativity: A Study in the Thoughts of S.L Jaki* (1996), is therefore a theological response as he himself states it:

> This work then involves a synthesis of Jaki's insights into the relation between Christian faith in God the Creator and modern Science. Since the present work is essentially theological, the focus will be on the theological aspects of Jaki's work on science and its consequences for theological reflection.[13]

This is why this chapter will take a philosophical approach in expounding on the vast resources of Jaki on the birth of science in Western Christian Civilisation especially as an intellectual operation that requires both the predominance of sound reason and intellectual honesty.[14] Jaki himself repeatedly confirms the need of a robust philosophy and argues that whoever wants to avoid metaphysical questions should not even attempt to speak. This is important for us because as we talk of science and its Christian origin, we shall be dealing with metaphysical issues that relate to Christian faith. The metaphysical question concerns anyone capable of using reason because, the mind itself dictates the acknowledgements of realities that transcend itself. Asking metaphysical question is therefore not optional but a necessary process of the mind. What is optional is the attempt to give answers to metaphysical questions raised that requires a special cultivation of the mind and intellectual honesty.

Thomas D. Watts has already pointed out the uniqueness of Jaki's works and how it can be used rationally to refute modern atheism. I can say that Watts gives a good indication to the real issue that should occupy a philosopher even when he does not treat directly the rationality of the science and Christian faith. Stephen M. Barr, in "Science Seeking Understanding" (2009), raised the need for a robust philosophy (besides a robust theology) in understanding the relationship of science and faith in Jaki's works. This gives credibility to the use of philosophy and serves for an impressive recommendation that we want to take up and develop in this exposition.

---

13 *Ibidem.*, p. 7.

14 Cf. S. Trasancos, *Science was Born of Christianity: The Teachings of Fr. Stanley S. Jaki*, The Habitation of Chimham, Titusville (FL) 2014. This book is written using theological and ecumenical arguments to promote the thinking of Jaki on the viable birth of science in the Christian circle.

Philosophically speaking, such a relationship is treated only generically in *Science and Faith in Stanley Jaki. Proceedings of the Summer Course* (2017), edited by Guerra Menédez *et alii*. I would like to go into the specificity of that which defines this relationship. I would like to acknowledge that Antonio Giostra in his book, *Stanley Jaki: Science and Faith in a Realist Perspective* (2019), has made a good attempt to reconstruct Jaki philosophical thoughts looking at it from a realist point of view, treating specific philosophical arguments to justify his position that Jaki, can be classified among philosophers as a realist. It is a collection of essays that deals with Jaki's specific philosophical ideas. The work of Giostra is a good philosophical tool; it however, does not specifically talk about the rationality of science and Christian faith around which I intend to reconstruct Jaki's philosophy and try to understand him more deeply. Man is defined by what he thinks and knows about the things that surround him and what he believes is the cause of what surrounds him. This is the reason for which science and religion will always remain at the centre of every serious discussion about the cosmos and its content.

I will treat the problem of rationality and method because they are two very important tools that are used in understanding faith and science. Jaki uses very specific methods to arrive at the conclusion that science could not be viably born elsewhere except in the Christian West. He insists that reason dictates some respect for facts that are historical that cannot be ignored without falling into methodological errors and therefore ceasing to be scientific. I will also treat intellectual honesty as a methodological and rational requirement.

I will discuss the concept of science generally and in the understanding of Jaki because when he talks of science, he wants it to always mean "exact science" which is concerned with the measurement of quantities and things in motion. We shall see the strength of his understanding of science in his arguments for the singular viable birth of science in the Christian West. Whenever we shall be talking about science according to Jaki, we shall always be referring to exact science unless it is specifically stated otherwise. Our first chapter will also treat the limits of science in relation to other disciplines because it contributes to the debate on the relationship between Christian faith and modern science. We shall demonstrate how the respect of limits lead to a fruitful dialogue between Christian faith and science.

Another important aspect that we shall discuss in this chapter will be the origin of science, which tackles the core of Jaki's arguments in which he shows the contributions of Christian philosophers, Christian doctrine and the Catholic Church in the viable birth of science. We shall also look at some oppositions to the view that science was born of Christianity and the reasons for such views.

This chapter will also deal with the major reason raised by Jaki as to why science could not be born in other civilisations despite the fact that there were elements of scientific capabilities among them before the Christian times. Jaki sustains that pantheism was a major obstacle in the viable birth of science in many civilisations. I will also present the views of other scholars on the arguments of Jaki, assess them philosophically and conclude the chapter. I have read many books in the course of my academic formation that most of the times have good titles that do not correspond to their contents. You will find in this book a constancy on the line of rationality of science and Christian faith and whenever I will use the term religion, it will really be returning to the Christian religion unless it is necessary to make a reference to other religions.

1.2 The Problem of Method and Rationality

Rationality of science and faith requires of itself the desire not to supplant each other but to ultimately seek its growth with some common sense. It is this very common sense that can promote harmonic relationship between science and faith because it should ordinarily be possessed by anyone who engages in serious discourse about the same. In order for science to be true, it is not sufficient that it be rigorous, it must depart from good sense in order to return to good sense.[15] This is the sense we want to sustain in this exposure as the rationality with all that it bears on the good use of reason. Without adequate reason, scientific endeavours are void, blind and dangerous not only to faith but also to humanity and consequently to the scientists. Jaki, in *the Saviour of Science* (2000), argues that some philosophical systems like Darwinism, to mention but just only one, fail to

15 Cf. S. L. Jaki, *Scientist and Catholic: Pierre Duhem*, Christendom Press, Front Royal (VA) 1991, 271.

provide a balanced use of reason in understanding nature.[16] To take evolution as the answer to the origin is to deny facts that history has proven beyond the theory that has no proof of a beginning except the breaking of an infinite backward return in history. It is even more challenging that Darwin himself did not define what species are with all the complexities that it now bears. Human reason accepts that there are things that it cannot fathom or explain which are reasonable and undeniable. Here, returns the demand for common sense that realities cannot be reduced just to only one thing. Having common sense is actually taking seriously common realities that we sometimes ignore because of the sophistication of the world. The reality of bodies is complex and good science seeks to understand complex parts in simple forms. At the end, simplicity is the point of departure of good science. Common sense, which is unique to humans, acts as a guide to all other senses. The first perception that every sense gathers is what makes it common sense. We ought to bear in mind too that it is unique to humans who at the end can do science and relate to metaphysical entities helped by their faith.

In understanding the relationship between disciplines, Jaki argues that philosophy, theology and science should not ignore each other but remain within their own competence. The universal view by philosophy is like a ladder to rise to a more holistic vision of reality using the mind without having to deny the contributions of other disciplines. Jaki argues thus: "Just as scientist cannot ignore philosophy, philosophers and theologians can only at their peril ignore the fact that quantitatively verifiable parameters as such lie outside their competence."[17] It is apparent here that the limits of science, theology, and philosophy are defined by the formal objects of their engagement and the methods employed in this engagement. For Jaki, experimentally verifiable quantities represent a basic demarcation line between the sciences and other forms of reasoned discourse such as philosophy and theology and therefore faith.[18] The issue of demarcation comes strongly also because science developed eventually in history as a separate

16 Id., *The Saviour of Science,* 25.

17 Id., *The Limits of a Limitless Science and other Essays,* ISI books, Wilmington (DE) 2000, 14.

18 Cf. *Ibidem.*

discipline and as it develops and becomes more specialised, scientists and theologians begin to define themselves within the limits of their methods, tools and competence. This is also necessary especially when Jaki argues that science was born within the Christian contexts, nurtured by Christian doctrines. Eventually the time came that this separation should be reasonably considered.

Jaki avoids the use of the word science to refer to philosophy and theology and it is here that his conceptual separation is evident. For him science is separate from theology and philosophy even when they can look at each other in the eye.[19] For him there can be no complementarities, in the strictest scientific sense between science and religion even when science was born within the context of the Christian faith. There could however be a possibility of some fruitful dialogue between the two since they have a close relationship in their origins.

Religion has answers to puzzles of science only when scientists have the eye to see that "beyond science revealed religion looms large as its saving grace."[20] Talking about intellectual honesty, Jaki adds on to it courage which is very important in promoting a good relationship between any two disciplines. Courage is a virtue that lies between cowardice and foolhardiness. One should have the courage to defend prudently the truth even when it may be uncomfortable and unpopular. Sometimes scientist defend their theories simply because they have become popular over time or because they belong to a particular scientific community or club. This blinds them to the fact that quantities are infinitely measurable to the smallest details as well as to the largest extension. This applies to myopic theologians too, who see science as a threat to the deposit of the faith and therefore, should be resisted.

Methodological limits should always be adhered to in order to avoid unnecessary conflicts. For Jaki it does not make sense for the scientist to speak about purpose in life process. This is because purpose can neither be measured nor experimented. The denial of it follows the same line because there is apparently no authority to do it on the part of the scientist. Jaki

---

19 Cf. Id., *The Impassable Divide,* Real View Books, New Hope (KY) 2008.
20 Id., *The Limits of a Limitless Science and Other Essays,* 30.

argues correctly that purpose is a matter of philosophy. Philosophy employs the rigorous use of reason to make legitimate connections where empirical method is unable to. It is only in this sense that there is rationality in scientific endeavours. It is therefore arguable as Jaki puts it that "in the absence of philosophy science can gain no better hold of the magnificent process of evolution leading from amoebas to man than one can grasp the morning mist."[21] This phrase when well understood tells us all about the role of philosophy in critically showing connections between brute facts within the holistic system of the universe and all that it contains to unfold in it.

1.3 The Concept of Science

Jaki's understanding of science developed over time to a more specific and demarcated concept purging it with time of the elements that do not belong to science. In *The Relevant of Physics* (1966), Jaki describes western science thus: "Western science rests on two foundations: the formal, ordered system of knowledge and the establishment of causal relationship by systematic experimentation."[22] Even when Jaki, [because of the zeal for physics] rejected this rich definition, he would return to it in his later works. The philosophical element in the above description of western science is what science has been and what it should be. Other definitions seem specificity of minute details of science as defined by physics, biology, chemistry and other applied sciences. Science is a rigorous discourse related to the acquisition of knowledge.

I will not be defining science here but I will only refer to science the way Jaki understands it. In the development of his ideas, Jaki reached a point of using the term "exact science" to exclusively define science as only relating to quantities and their measurements. He uses the term 'exact science' in the evolution of his thoughts to distinguish experimental science from other studies that for him are not worth calling sciences because of their inability to predict results accurately. This was a cautious approach that was open to explore what the reality of the word science has come to mean. Let us examine his arguments.

---

21 *Ibidem.*, p. 124.

22 Id., *The Relevance of Physics,* Scottish Academic Press, Edinburg 1992, 4.

Jaki defines science as "the quantitative study of the quantitative aspects of things in motion."[23] He goes ahead to explain that the definition was meant to save science and the humanities from mutual encroachment and to avoid methodological error always committed. According to Jaki, what many people see as conflicts between science and faith or any other disciplines that have conflict, with is because science is always not well defined and forced to do what it is unable to do. According to Jaki, science has nothing, for example, to offer to religion as a thing directly coming from it, because it has no method to handle religious things.[24] This however does not mean Science and Religion cannot come into dialogue. They actually need dialogue to avoid the encroachment of boundaries and limits because they both need reason to communicate their contents.

Jaki presents very clearly throughout his professional carrier that science cannot provide a definitive answer at any point in time to all questions regarding nature and existence. There is always something new that science cannot explain and can explain after sometimes in a better way. Science is therefore all the time incomplete in all its methods and tools. It answers limited questions about the cosmos and only those within its competence. This is a limitation for science but also its strength because; it is what compels it to move into new horizons of discoveries. Consumers of scientific discoveries ought to remember always that discovery is not creating something which was not but acquiring knowledge of something that has always been that was not known. This helps to appreciate the fact that science works on a given and when a given is not known, it does not mean that it is not. Epistemologically, science forms part of the vast capacity of man's ability to know and remains always partial and yet within the epistemic potentiality of man. Jaki argues that science should have a purpose and should be driven by the search of the truth even when it is not the case most times. If one were to ask Jaki what science is, he would most probably reply: exact science is physics that provokes cosmological and metaphysical questions whose answers go beyond their competence and tools. This then opens a window for us to investigate into the limits of science.

---

23 *Ibidem.*, p. 67.

24 For an elaborate analysis of Jaki's concept of science. Cf. P. HAFFNER, *Creation and Scientific Creativity*, 26–52.

1.3.1 Limits of Science

When scientists formulate laws about regularities in nature, they reach a level where their method is no longer helpful and they should employ the philosopher's mind and method to proceed. For instance, the astronomer cannot answer the question why the orbital bodies move round the sun. The scientist cannot tell why there are regularities in nature from which laws of nature can be derived or even why should there be a law of nature and not nothing at all. Science is limited and actually all other disciplines are limited by many factors. Haffner interpreting Jaki asserts: "Science is an intellectual activity involving man, his instruments and the cosmos, none of which is provided by science."[25] Science therefore to merit its name should be guided by intellect that seeks to investigate into the cosmos as given. This means that science begins at the point in which scientists recognise that there is something that merits investigation. Science is therefore, limited as long as man is and as long as it deals with what is not caused by scientists. This means that scientists are limited to the materials things at their deposition and limited by their very nature as humans. There are therefore many limits that scholars should have the ability to recognise in order to respect them.

It is therefore important that when such a limit is reached, scientists should be humble to accept their limits and leave the construction of knowledge to the competent other. The question of limits has been the concern of many philosophers with Jaki being no exception; in fact, Jaki dedicated a whole book on the limits of a limitless science. Richard Connell asserts that "uniquely human character is the desire to solve problems that are purely theoretical and of interest for their own sake."[26] The interest here may be scientific, metaphysical, philosophical or even religious depending on the formation and orientation of each scholar. "Thus, in man, problem solving ability refers not exclusively but primarily to theoretical activities

---

25 P. Haffner, *Creation and Scientific Creativity*, 42. For a general overview of the thoughts of Jaki on the limits of science, Cf. Id., "The Limits of Science in Stanley Jaki," in G. Menéndez – R. Pascual – A. Colombo, (eds.), *Science and Faith in Stanley Jaki. Proceedings of the Summer Course*, IF Press, Rome 2017, 37–56.

26 R. J. Connell, *From Observables to Unobservable in Science and Philosophy*, University Press of America, New York, 2000, 128.

that are legitimately goals in themselves, in that way man is most fundamentally distinguished from other species."[27] First of all, science is born from problems that must be solved because it affects the human person and because it is limited, sometimes it solves one problem and another one emerges or in solving one problem it creates another. According to Karl Popper, that is how science progresses, solving problems, hypothesising solutions, experimentation and facing more problems. Behind and beyond every scientific problem, there are several other problems that are left out in order to pursue the answer to the problem at hand. This is also an element that makes science limited in the sense that it does not provide answers to all questions as most of the time we think. It provides answers to questions that concern particular line of investigation. When one is mining gold, he is likely going to throw away into the debris of his mind precious minerals that he does not know because they are not gold. In the same way, in the debris of science there are many unanswered questions that science either cannot answer or it has not become a matter of concern that provokes an investigation. Man in his desire to solve problems meets his limits and honestly realises that much of reality lies beyond his reach. He then either reaches them by his faith guided by reason and revelation or by refinement of his 'tools' to dig more deeply into knowable and observable reality that he will never exhaust because of the shortness of his life, the limits of his tools and the infinite problems he raises by trying to solve just a single one. At the end of his endeavour, he does not solve a single problem but gives the most plausible explanation of it at that given time and as long as that explanation lasts. If you are a fanatic believer in science, you should be getting scared by now to face the reality of your believe in the capacity of human ingenuity.

The importance of knowing the limits of scientific methods is a pure necessity for scientists. Jaki argues that the truth has its intrinsic value regardless of application or action and controls in the end all pursuits scientific or others.[28] What therefore remains in the pursuit of any knowledge is the truth of its content. We recall here that the object of faith is the

---

27 *Ibidem.*, p.129.

28 Cf. S. L. JAKI, *The Limits of a Limitless Science and other Essays,* vii.

ultimate truth and therefore if science seeks some truths based on facts, it has a relationship with faith. He continues to argue that, as far as the scientific method is concerned, there are many limits to science and yet in applicability of quantitative measurement, science is limitless. As long as science seeks to investigate into substance, there are endless possibilities of discoveries. It is also limitless in the sense that scientists will not go on holiday because they have understood nature. Scientists will always understand only a very small part of the totality and very limited aspects of that very small part. "What science cannot do today may well be done tomorrow or the day after."[29] This is apparent because what is known or unknown now is affected by many factors ranging from the scientists' intellectual ability and the effectiveness of the tools in his hands.

Any quantitative or empirically stated problem can have its answer in science even if that science is not yet known. In that sense there is nothing beyond science as long as it remains within the realm of quantity and measurement but as soon as one faces basic questions about existence and knowledge, science is limited and cannot respond adequately to those questions. One has to go beyond science to arrive at science itself.[30] By saying that one has to go beyond exact science to arrive at science itself, Jaki actually affirms the unifying role of philosophy as a science that studies underlying principles which relates to knowing.

When dealing with limits and boundaries of disciplines we ought to avoid extremes by defining our terms. Jaki in his early works borders on some form of reductionism by attempting to reduce science to the laboratory but avoids the claws of reductionism by not making it the only acceptable body of knowledge. Science has in the course of history acquired several meanings. Commenting on theology and philosophy as science long defined by St. Thomas and St. Augustine, Jaki asserts that "one wonders whether they would insist on that name nowadays when the word science has a very special meaning attached. That special meaning is attached to the laboratory and the kind of work done there."[31] It is however clear that this is the meaning our author wants to attach to science and not simply

29 *Ibidem.*, p. 103.
30 Cf. *Ibidem.*, p. 104.
31 *Ibidem.*, p. 2.

that the understanding of science has shifted from what it has been. We can always talk comfortably about science as experimental and discursive. This is because science in its very original etymological form has an epistemological dimension. The development of the physical sciences or exact science as Jaki prefers to call it does not make theology and philosophy less scientific but makes it different sciences. It would be interesting to consider the history of the word science and how it has developed into its current use. We are however, methodologically, more accurate if we make known right from the beginning what we mean when we use a certain word. This ability is what philosophy does in its defining role.

The term 'science' etymologically, comes from the Latin '*scio,*' which means 'to know' and in a more rigorous way, from the Greek word '*episteme.*' Science in the epistemological sense is the search for knowledge and above all for the truth. Experimental science is that which is concerned with knowing as ordered to doing. This type of science does not concern itself with the study of the universals as metaphysics would do, but is concerned with specific and particular reality based on empirical evidences and experimentations, that finds meaning in a universal context. It was not until 2005 that Jaki, in the "Forward" of the reprint of *The Road of Science and the Ways to God* (2005), comes out clearly about the epistemological aspect of science giving its dominance to philosophy and theology:

> Only those armed with philosophy can take comfort in the fact that theological and philosophical presuppositions played, however, an important role in that part of scientific enterprise which is the raising of an epistemological scaffolding around the strictly quantitative, or mathematical part of exact science.[32]

Such an affirmation shows a very high level of intellectual honesty and development of ideas and may be more emphatic not just as a mere scaffold but a foundation on which science stands if it is to be communicated as knowledge to a third party.[33] Probably more important is that failure to comprehend that a scientific theory does not constitute its irrelevance to

---

32 S. L. Jaki, *The Roads to Science and the Ways to God*, Real View Books, Port Huron (MI) 2005, vii.

33 Jaki notes that every publication is an evolution of the author's mind and clarity in one's philosophy develops overtime.

faith and religion but an invitation to take on another level of investigation in search of truth.

For Jaki, in relation to accurate observations and prediction science is synonymous with measurement, which are accurate because they can be expressed in numbers.[34] The purpose of restricting science to measurement by Jaki is to show how incompetent science can be in dealing with non-quantitative entities qualities abandoned already by Galileo in his mathematization of the universe. The fact that non quantitative properties cannot be expressed mathematically does not mean, however, that they do not exist but rather that they do not form part of what physical science can competently deal with. "Science is competent whenever the object of investigation offers a quantitatively determinable aspect."[35] Whatever is not measurable should not be the preoccupation of physical science. To go into discourses beyond the laboratory is to make an illicit jump. Science therefore has two understandings of limits: unlimited in as far as measurement of physical objects are concerned and limited in as far as it cannot establish by its methods the why and the what of things. According to Jaki, science cannot even define a thing except describing it. It is clearly better, according to Jaki to take another attitude and acknowledge that there are some basic limitations to a limitless science. Those limits appear as soon as questions arise that cannot be put in a quantitative form and therefore cannot be given quantitative answers to be tested in a laboratory.[36] These questions are definitely legitimate and can only be answered by the metaphysician, theologian and the philosopher who doubles as the arbitrator in the game. Scientists go wrong when they begin to think they should provide answers to every question going beyond their limits and method. This temptation is great in a world that thinks physical science is omnipotent, when the author of that science is contingent. I think this is a defeat of good reason and a sin against logic.

When we talk about Science we must be immediately talking about objects and their measurements. Positive Science, like any other science, cannot operate out of nothing and from nothing and actually, when science

---

34 Cf. Id., *The Limits of a Limitless Science and other Essays*, 2.

35 *Ibidem*.

36 Cf. *Ibidem*., p. 11.

begins to measure, it measures something whose existence he is unable to explain using his methods and tools. Precisely stated, positive science deals with a given. We rightly talk of scientific discoveries because science only discovers what is in existence but unknown at the time because of the limits of our capacities. Even scientific inventions are a combination of existing things and their rules to do something or answer to a question never dealt with before. "Science merely discovers new aspects in the vast gamut of material existence."[37] These new aspects are only new as long as it has not been known before and may still be known in a better way after. Scientific discoveries are therefore most of the times the overcoming of the ignorance of the physical world.

To realise the limits of science and any other discipline, there is need for intellectual honesty that necessitates two very important elements: intelligence and wisdom. This refers to what we know now and what we potentially can know within the span of our lives. It also invites us to consider the medium through which we can communicate our knowledge within the gigantic wealth of various aspects of knowledge. But also, the acceptance that our minds are capable of knowing only up to a certain level by its very nature and training. Jaki exposes throughout his work using questions; attitudes that point clearly to some elements of intellectual dishonesty. Referring to Newton refusing to comprehend the Milky Way, Jaki says, "Is not blindness to the obvious the strongest hint about some wound in the human intellect, and should it not be especially absent in an intellect which is almost superhuman?"[38] Even the smartest mind capable of great discoveries like that of Newton has its own shadow that is determined and conditioned by the possession of its mode of existence that he has no control over; but also because scientific investigations are geared towards the most urgent need that makes an investigator blind, sometimes to even obvious facts. If a scientist is to achieve progress, he must be selective to the problems he tackles without denying the fact that there are exist other problems that demand answers too. Tackling dishonesty in Kant as a sin he says: "But is not something sinful at play when the human intellect

---

37 *Ibidem.*, p. 12.
38 Id., *The Saviour of Science,* 103.

chooses to see certain things and to blind itself to others?"[39] Referring to Darwinists dishonesty contradicted by the function of their own eyes he says: "Why would it then be far-fetched and improper to see in this fearful paradox of illogicality a retribution from on high for a self-imposed myopia approaching voluntary blindness?"[40] It is so true that scientists sometimes would not want to recognise the so obvious because they are either prejudiced or because it serves as a defence mechanism for holding and sustaining a scientific idea at all cost. This makes them drift away from exact science into ideology and politics of science. Referring to Darwinists diluting philosophy into a comedy he says: "Nothing can reveal more poignantly the depth to which man can fall than that very high level at which he is ready to take sickness for health and vice versa."[41] These are strong expressions of disappointment in a selective use of the intellect and reason that seem to neither benefit science nor faith. The evidences of this kind of attitude and dishonesty are not lacking in our modern time in which being objective is a choice and being faithful to intellectual rigour is a burden to be left for a few. We would like to conclude in the words of Jaki: "of genuine intellectual humility one can never have enough."[42] There is always the temptation to postulate and speculate beyond empirical evidences which is itself the quality of the mind not satisfied with the bare minimum, but the same intellect dictates that one should remain within the limits of his competence that constitute that genuine intellectual humility greatly desired. We already know that at the social level, humility is already taken for a weakness and it must be very scaring at the intellectual level with the risk of masking in humility only by words when deep down such a humility is not genuine. To be intellectually honest and genuine does not mean being weak or defeated but giving opportunities to other possibilities in the understanding of the cosmos and its content.

39 *Ibidem.*, p. 105.
40 *Ibidem.*, p. 148.
41 *Ibidem.*, p. 164.
42 Id., *The Relevance of Physics*, 404.

1.3.2 The Origin of Science

Jaki in studying the history of science goes a long way to demonstrate that sometimes a scientific research is impeded because it loses the science by which it should proceed because of stereotypes. He discusses the reasons for which science was not viably born in ancient Greece, Egypt, India, China and other ancient civilisations. Jaki acknowledges the fact that there were elements of science in these civilisations but that they could not constitute a viable birth that singularly occurred in the Christian West. We shall discuss in this section some of the reasons for the failure of science in some selected ancient civilisations to demonstrate the arguments of Jaki for their failure.

1.3.2.1 Stillbirth of Science in India

According to Jaki, India could not give a viable birth to science because it lacked the philosophical foundation to liberate itself from Buddhism, which is fundamentally some form of pantheism. Such a standpoint suffocated science other than giving it a viable birth because it could not permit curiosity to delve into nature to establish its truth content. Among the Hindus, science failed because they could not get out of the enslaving cycle of birth and rebirth because their god formed part of their nature and thus leading to a resignation to leave things as they are because they will always return to the start point and run its course again.[43]

Jaki argues in relation to eternal return, "on the contrary, the fearful prospect of never-ending cycles filled one with distrust and hostility towards the world of senses."[44] The fear of the world of senses logically tantamounted to fear of delving into scientific research and thus impeding the birth of science among the Indians. Jaki argues that even when it occurs in India first, the decimal system or high level of craftsmanship, it does not add up to science. This is because according to him, "the very soul of science consists in theoretical generalisation leading to the formulation of quantitative laws and systems of laws."[45] The failure to account for the development of

---

43 Cf. S. L. Jaki, *The Saviour of Science*, 9–48. In these pages Jaki discusses the still birth of science in various ancient cultures and points out why they would not give viable birth to science.

44 Id., *Science and Creation,* Real View books, New Hope (KY) 2016, 11.

45 *Ibidem.*, p. 16.

science in India is also due to the lack of precision in their dating system. An argument that we can term as lack of sufficient evidence. Jaki substantiate thus, "In trying to put scientific root of Indian culture in evidence, the modern Indian scholar cannot help being frustrated by the uncertainty of the dating of what is best in ancient Hindu scientific heritage."[46] This makes it difficult to refute the fact that science could only be viably born in the Christian West because there is no proper scientific dating in India that indicates a scientific revolution comparable to that which occurred in the Christian West. Therefore, a systematic, sustainable science was not born India despite their ability in the use of numbers and other skills that showed the path to good science that presumably died before it could constitute a viable birth.

1.3.2.2 Stillbirth of Science in Egypt

According to Jaki science failed to have a viable birth in Egypt, even when the construction of the pyramids demonstrated high architectural style and genius. Egypt lacked the systematization required of a viable science that is sustainable. Ancient Egypt had "several impressive but disconnected insights against a vast background of practical skills and grossly erroneous beliefs." For a viable birth of science, there is need to have some consistency that leads to formulization of general principles which ancient Egypt did not have. For science to thrive, individual skills needed to be universalised and codified into a form usable by any one capable of it. Erroneous beliefs too, played a great role in blocking the viable birth of science among the Egyptians; animals and insects were considered sacred as manifested in various personifications of them in the *sedja, ammita* and *seba* to mention but only three.

Mathematic, which is a fundamental requirement of science, did not take root in Egypt because the Egyptians were overwhelmed by bookkeeping. According to Jaki "they failed to reach the level of using explicitly the simplest mathematical operators, the signs of addition and subtraction."[47] The failure here is to see the need for a theoretical formulation needed for

46 *Ibidem.*, p. 22.
47 *Ibidem.*, p. 69.

a viable birth of science. Science, being a quantitative study, needed clear mathematical system to express its formulae, which eventually would help in the expression of generalised scientific findings.

Astronomy failed too, because the Egyptians did not have a stable and clear concept of time. They had a ritualistic concept of time that was calculated to fit their lunar festival that plunged them into recurring cycles. This did not give the Egyptian scientists the liberty to study scientifically the movement of the planet even when they noticed the difference in the course of the year of day light time and night. This ritualistic mentality did not help in the line of medicine too because the Egyptians mixed up medicine with magic and sicknesses were attributed to evil spirits. For them God formed part of nature and had no commanding position, falling back to ancient problem of pantheism and all its consequences.

Science therefore, could not be viably born in ancient Egypt according to Jaki because the Egyptians lacked the mathematics required for formulating scientific generalisations. Their worldview was mixed up with myth, magic and eternal return that made the Egyptian scientist caught up in the wheels of survival and not advancing in scientific discoveries. If science is a systematic rigorous study whose advancement is promoted by constant experiments, such a milieu would not usher science into the eminent position that it enjoys now.

1.3.2.3 Stillbirth of Science in Ancient Mesopotamia

It is an historical fact that Mesopotamia is taken as one of the greatest ancient civilisations in which human beings learned to live together, build cities, with Ur being the oldest city in the history of humanity, and developed harmonious living. Jaki argues that even in the great Mesopotamia, science could not be viably born.

> Cultural life in ancient Mesopotamia had seen only one creative phase which coincides with the reign of the Hammurabi dynasty in Babylon during the nineteenth century B.C. The same period also represented the high-water mark of the political ascendancy of Babylon, and for this reason ancient Mesopotamia culture is often spoken of as, Babylonia, especially when its scientific components are considered.[48]

---

48 *Ibidem.*, p. 86.

Basing on the fact that I am considering the birth of science, I will be referring to Babylonia and not Mesopotamia. It is an undeniable fact that Babylon had at a very early time scientific innovations like the construction of the ziggurats that demanded expatriate technicalities. Jaki notes that weather conditions and natural calamities were not friendly to the Mesopotamians and were constantly reduced to subsistence, hence impeding their scientific and technological development. The ziggurats, according to Jaki, showed some elements of science. "The planning, building, craftsmanship and practical knowhow extending over wide areas certainly contained prescientific elements."[49] Various skills could be identified in the construction of the ziggurats including the knowledge of geometry, chemistry and metal works. This however, did not contribute for the viable birth of science because according to Jaki, the Mesopotamians could not develop mathematics to serve the purpose that could sustain science. He argues that "in other words the fate of mathematics in ancient Mesopotamia is more than a problem of progress and stagnation. It is rather one of the several historical instances illustrating the repeated inability of scientific thinking to reach a state in which progress becomes self-sustaining."[50] In Jaki's thinking, the Babylonian mathematics lacked the realism necessary for the sustenance of science. They saw numbers everywhere in the Pythagorean's style but did not use them for concrete and practical scientific generalisation that science required for its viability.[51]

In the field of geometry, which has more in common to physical realities, the Babylonian science was not much better. According to Jaki, "Babylonian's geometry remained rudimentary in its dealing with practical determination of areas and volumes of elementary shape."[52] It can be argued that as long as exact science is about the measurement of quantitative elements of things in motion, the failure to express length, width, height, and area in concrete physical terms constituted the failure of science. This is one of the points that Jaki insists on, that science cannot be done a priori and therefore, numbers must reflect concrete realities in nature.

49 *Ibidem.*
50 *Ibidem.*, p. 97.
51 Cf. *Ibidem.*
52 *Ibidem.*, pp. 88–89.

In astronomy, the Babylonians were impressed by phenomena and were unable to form generalisations that would make predictability accurately possible. Focusing on phenomenon definitely made them miss the ontological foundation of reality and therefore missing too the realistic approach to physical reality available for investigation. In fact, according to Jaki, their magical background would not help them initiate complex astronomical concepts. Jaki affirms, that, "the non-scientific nature of the Babylonian astronomy is strikingly evident from the fact that it has never developed even tentatively a geometrical, or a mathematical model of the system of the planets."[53] This is indicative of the failure of science because planetary systems are empirical in the sense that they affect human by visible changes in seasons and time reckoning that should have provoked curiosity and interest to dig deeper into its content and mode of operation.

The failure of science in Babylon can also be attributed to their cultural cocoon that made them shun all Greek ways and could not borrow the mechanical planetary system that was already highly developed among the Greeks. In fact, they were more open to India than to Greece. It can be argued that for science to be viably born, it needed to be open. This openness would not exclude borrowing advanced scientific discoveries from neighbours and scrutiny from other civilisations. Science eventually develops by checks and balances, by being open to scrutiny by the scientific community either for confirmation or proving the validity of particular scientific findings. This has enriched and promoted science in the course of history. According to Jaki, this lack of openness made Babylonian astronomy "unknown to the Western world until the late nineteenth century."[54]

Jaki warns that the vast evidences presented by many historians of science about the astronomic advancement of Babylon may be misleading when not critically scrutinised. This was because they were essentially astrologers rather than astronomers and their motivation were not science. Jaki affirms that:

> Babylonian astronomy was not interested in the facts of heaven as scientific data. Its interest lay exclusively in providing a fairly correct lunar calendar, completely subordinated to the service of religious rites. These in turn were expressions of an

53 *Ibidem.*
54 *Ibidem.*, p. 90.

> animistic and cyclic worldview that provided no fertile soil for the development of scientific enterprise.[55]

We have already seen that, in most cases of the failure of science, there was the failure to break away from the mythical and religious explanation of the content of nature escalated by the cyclic worldview that gave no motivation for scientific venture. The Babylonians did not qualify for a viable birth as their search was influenced by religion and not subjected to the rigorous rational scrutiny that the medieval Christians were capable of. In other words, the rational rigour required for science, accruing from the rationality of the universe was not developed. According to Jaki, "the skilful calculators of Babylonians' lunar theory were, it seems primarily astrologers, magi, and soothsayers."[56] I must assert that this is one of the rare occasions that Jaki uses a moderate term most likely owing to the fact that he is dealing with an ancient civilisation whose history is remote and difficult to reconstruct but also whose dynamism would not permit of a generalisation.

In the area of cosmology, the Babylonians also failed, because the cosmology presented in the *Emma Elish* mythology was an inappropriate tool for scientific revolution. According to our author, their cosmology was full of superstition and animism with violent succession of the gods who had *Marduk* as their commander. Interesting to note is that he (Marduk) fashioned the world out of the blood of the other gods that he killed. The concept of creation out of nothing by an independent creator was not considered by the ancient Babylonians.[57] Jaki argues therefore that "in Babylon the succession of years did not mean the accumulation of building blocks with continuous constructive potential. They rather meant the transition from one world period into another separated by the discontinuous chaotic condition."[58] Science definitely needed to evolve consistently for a viable birth and this progress should also be sustainable and Babylon lacked this quality. I would like to conclude this section with the argument of Jaki thus:

55 *Ibidem.*, p. 91.
56 *Ibidem.*
57 Cf. *Ibidem.*, pp. 93–97.
58 *Ibidem.*, p. 97.

> The animistic, cyclic worldview made it, however, impossible for them to realize that to influence or control nature one had to be able to predict accurately its future course. [...] As a result, the mastery of science could not become a proud feature of the culture of the land on which ziggurats cast their sombre omen.[59]

The failure in mathematics, astronomy, cosmology and the failure to break away from pure religious explanations of natural reality, and the failure to scientifically predict results from the operations of nature, therefore resulted into the failure of a viable birth of science in Mesopotamia.

1.3.2.4 Stillbirth of Science in China

The Chinese, according to Jaki, harboured the organismic concept of the universe and this state of mind could not allow a viable birth of science. He argues that "[when] Tung Chung-Shu succeeded in making Confucianism the official state doctrine in 136 B.C., the emphasis on organismic consideration and their close ties with cyclic patterns became even more evident."[60] The Taoist system based on Yin and Yang did not help the Chinese either since it ruled out purpose from their investigations and the mastery of nature lead to abandonment.[61] It is needless to say that, with this kind of mind set, scientific investigations into the details of nature would not be a motivation because the cosmos appeared considerable to investigate into. Another factor that affected Chinese science was cosmic randomness cultivated among its sages. This according to Jaki made them not pay attention to "measurable, quantitative aspects of events occurring closely in time." This randomness could not inspire them the desire for a systematic observation and probable invention. "The universe was not therefore scientifically investigable, as scientific investigations of nature makes sense only if nature is consistent in its operations."[62] Unlike the Christian West, the Chinese did not have the concept of a rational creator and purpose. This

---

59 *Ibidem.*, p. 101.

60 *Ibidem.*, p. 27.

61 Cf. *Ibidem.*, p. 30. The system of ethics, education, and statesmanship, taught by Confucius and his disciples, stressing love for humanity, ancestor worship, reverence for parents, and harmony in thought and conduct. While Taoist is the philosophical system evolved by Lao-tzu and Chuang-tzu, advocating a life of complete simplicity and harmony.

62 S. L. Jaki, *Impassable Divide*, Real View Books, New Hope (KY) 2008, 39–40.

made the universe a product of irrational-random-self-creation. One basic requirement for science is actually its rationality and if the universe is not rational, then viable science is not possible. Such a worldview therefore could not provoke the scientific revolution that the modern time enjoys.

1.3.2.5 Stillbirth of Science among the Muslim

Science failed among the Muslim and the Arabs, according to Jaki, because their scholars held silently a parallel worldview to what Islam and the Koran offered them. Besides, the Koran taught that the Will of Allah must prevail over all physical laws and holding a contrary view was considered apostasy. The Islamic philosophers never opposed openly the teachings of the Koran and their renounced scholar Averroes fell back to the eternal world of Aristotle. "Averroes was firmly convinced that psychological and social catastrophes of the worst kind would result from the untimely revelation of the esoteric meaning of religious prescription and teaching to the weak and the ignorant."[63] Accordingly, the weak and the ignorant should be kept at their level of ignorance as an act of charity to them since exposing them to higher knowledge would be risking both their lives and faith. Among the Muslim, therefore, the fear of contradicting the Koran suffocated science as efforts in scientific research were directed towards conformity with what their religion taught. Jaki argues: "Orthodox Muslim scholars did their utmost to undermine the notion of the universe operating along consistent laws, calling it a taint on Allah's absolute freedom to do whatever He wanted."[64] This did not promote disinterested thinking that is required for science. At the end, science among the Muslim was interpreted according to the Holy book impeding the viable birth that science needed.

1.3.2.6 Stillbirth of Science among the Greeks

According to Jaki, Greece is taken without question as the birthplace of science to the extent that no one even dares to investigate into the truth of that assertion. For him, "the phrase Greece the birth place of science has

63 H. Corbin, *History of Islamic Philosophy*, Kegan Paul International, London, 1993, 245.

64 S. L . Jaki, *The Saviour of Science*, 78.

been presented to modern man's eyes and dinned into his ears so often and so systematically that even if he is a historian of science, he may not have second thoughts about it."[65] Apart from the argument raised about the non-viable birth of science among Greeks, Jaki points out a serious problem of philosophical concern, which is the deliberate lack of critical thinking and curiosity to establish the truth and falsity of a popular opinion, simply because an unquestionable ideology has been built around it. Besides verification, confirmation of facts is the greatest commitment of science and the fact that it is not done in the tracing of its own history leaves a lot to be desired. Jaki gives a panoramic account of how Greek world view impeded the viable birth of Science beginning by taking into consideration the works of Plato.

According to Jaki, Plato's a priori and the Demiurge borrowed from Socrates' purpose and Aristotle's eternal world imprisoned science into an infinite gestation among the Greeks. According to him, "Aristotelian universe bespoke law and order but not a legislature."[66] This implies that Aristotle could not account for the author of the laws of nature that he ably demonstrated. This made him not to ascend to creator outside nature itself, which is a required step for an independent investigation into nature that constitutes science. This gap gave the universe its divinity in eternal return without a centre or beginning and with a movement from glory to destruction. Eternal recurrence is a cosmological view that impeded science because it did not give it the necessary impetus owing to the fact that, whatever is now could have occurred in a remote past and will reoccur in a long distant future. This did not promote the curiosity on the part of the scientists, because what is not known now could have been known before and somehow in the cycle it will come back, making scientific investigation a vain search. Consequently, there is nothing new and complacency sets in as a justification not to venture into further research. With this worldview, scientists would have no motivation to carry out research other than wait for it to return in its cycle. One would definitely wonder, "But if everything

65 Id., *The Origin of Science and the Science of its Origin,* 1. Cf. Id., *Patterns or Particles and Other Essays,* The Intercollegiate Studies Institute, Bryn Mawr (PA) 1995, 161–178.

66 Id., *The Road of Science and the Ways to God,* 25.

repeated itself in an infinite number of times, what was the point of making a major effort to improve one's lot?"[67] One's lot would logically mean what one's fate dictates upon him and the safest thing to do would be a wholesome acceptance of it.

The view of the cosmos as divine made scientific investigations even more difficult because the cosmos was seen as a living being. To investigate into the divine universe is tantamount to disturbing the gods as Epicurus would later on preach it.[68] The scientist therefore is inserted into a totality from which he cannot liberate himself and yet science demanded that the object of investigation should be something outside of the subject for which investigation is possible. If the cosmos is divine, then science has no tool to carry experiments with divinity. This impeded the birth of science in as much as the investigator did not go forward on facing unexplainable problem and sought explanations in some sort of divinity that formed part of nature. The same breakthrough that the Ionians had, seeking for explanations of nature outside Greek mythology, was also needed for science, that is investigating into nature and not leaving its explanation to myth or popular opinions.

Ancient Greece, according to Jaki, had many opportunities with individual scientists who however never got out of the cyclic trap set for them by Aristotle. It was the stoics who brought the Greeks closer than any other culture to formulating a viable science. "Stoicism marks the burning out of that flame of scientific creativity that was alive only for a few generations."[69] It is here that we get to appreciate what Jaki means when he talks of a viable birth of science that is proved, by time as flame that produces light and not just a flash. He adds, "No sooner had the human mind risen to such heights than it slumped back up itself."[70] For him the Greeks failed to make science grow by not making it broad and persistent for everyone to embrace because that is how science works and survives. The nature of science is the openness of its findings for the scrutiny of everyone. According to Jaki, the Greek civilisation was closed and limited to one's capacity to learn Greek before entering into their philosophical system. It was also characterised by

67 Id., *Patterns or Principles*, 167–158.
68 Cf. Id., *The Road of Science and the Ways to God*, 26.
69 *Ibidem.*, p. 33.
70 *Ibidem.*

more speculative studies than experimental adventure. Scientific openness however, most of the time remains in principle than practice because of the politics played by the scientific community. Rigorous science should at all times be separated from the activities of the scientist's clubs.

It is worth noting that Heraclitus' view of the cosmos as flux, which influenced many other Greek thinkers, was not conducive for the viable birth of science. If science is to be that rigorous, consistent investigation into matter and its content, stability is definitely a prerequisite. If things are in flux, then logically science is impossible, because the experiments carried out today will not be the same one that will be carried out tomorrow and in this way binding conclusion would be an illusion. It would also be impossible to establish some of the law of nature, which is the law on which science thrives. Some stability was therefore needed for science to become what we now have. Science advances by constantly examining already established answers for their accuracy and this would not be possible if everything is in flux. According to Jaki, therefore, science could not be viably born in ancient Greece and the biggest underlying factor that affected most ancient cultures was the pantheistic vision of the cosmos.

Etymologically, the term 'pantheism' is derived from two Greek words *pan,* which means all, and *theos* which means God – *pantheos*, therefore, meaning God in all things and also all things in god. Pantheism is the philosophical doctrine holding that there is a bit of god in everything. It has in history taken various forms, for example, God being equal to nature, God being contained in everything, God having power over everything and god as the vital force behind the operation of every phenomena. God is identified with the cosmos as one thing in the positive sense and in the negative sense pantheist also believe that there is nothing in the universe that exists outside God. The universe and God are two faces of the same thing.

Jaki takes the early discussion of the understanding of the universe as composed of the four elements: earth, fire, water and air, proposed by the Ionians philosophers to the time of Aristotle, as pantheism and extends this concept to modern pantheists like Spinoza.[71] We note that he uses

71 Cf. S. L. JAKI, *God and the Cosmologists,* Real View Books, Fraser (MI) 1998, 92.

pantheism in a generic manner, when discussing ancient cultures to refer to a belief in the eternity of the universe and the eternal return through reincarnation and rebirth. This is probably because he discussed in depth the ancient Greek's concept of the cosmos as an organism in the *Relevance of Physics* (1966).[72] Any culture that was unable to reach the knowledge of the cosmos as created and as outside God is generally taken by Jaki as pantheistic, a factor that blocked the development of science because it also considered the cosmos as a living organism.[73] Jaki describes a pantheistic God as one that has "never been the object of fear, let alone of hope, precisely because such a God is bordering on non-existence."[74] The pantheistic God has no identity since it is everywhere and in everything and Jaki identifies these elements when he discusses the fact that other civilisations could not give birth to a viable science because they had no clear path to transcend nature as God was in matter and matter in God.

Jaki argues that the Greeks never parted from the concept of nature as an organism, even Aristotle promoted what he calls the Socratic program in which "nature was to be understood as something like man himself – moving toward goals, striving towards the best possible arrangement, in short, acting like an organism."[75] Despite the fact that this concept contributed a lot to the understanding of nature as oriented towards an end, and therefore purposeful and rational, it did not help Aristotelian science to escape the pantheistic vision of the cosmos. The fundamental breakthrough of the Ionians seeking explanation of the universe outside the Greek's gods fell back to the explanation of the cosmos in itself. This was frustrating and self-defeating and could not cause the revolution science needed.

In the field of astronomy, Jaki argues that the Greeks failed too to take science to the next level that could have enjoyed the present scientific revolution, because "they did not advance from geometry to dynamics [...] Dynamics comes into picture when one analyses the motive force that drives the machine."[76] Indeed the Greeks failed to attribute to the mover the

72 Cf. Id., *The Relevance of Physics,* 3–51.
73 *Ibidem.*, p. 91.
74 Id., *The Saviour of Science,* 238.
75 Id., *The Relevance of Physics,* 43.
76 Id., *Patterns or Principles*, 166.

right concept. In this way they failed to define celestial motion outside the popular Aristotelian vision. We note that, if the Greeks failed, it was not because there were no technology or science among them, but it was precisely that they did not take it to the level that Galileo and Newton were able to. We have to then consider why Jaki thinks that science needed Christianity to take it to the next level that constituted the scientific revolution we are now proud of.

1.3.2.7 Reasons for the Viable Birth of Science in the Christian West

The failure of science in all ancient cultures and religions went hand in hand with the failure to arrive at a clear knowledge of God, who created the universe subject to scientific investigations through natural theology. Jaki defines natural theology as "the highest and deepest form of the exercise of a mind which, precisely because it knows truth, can be known to reflect, however modestly, the infinite rationality of an absolutely truthful God."[77] The failure of natural theology constituted the failure of science, according to Jaki, because first of all, it was the failure of reason to arrive to God through nature and secondly, the ancient cultures were unable to separate creation from its creator, one big step that could have taken them away from the snares of pantheism. Jaki affirms that "there can be no fundamental difference than the one between a world which is in no need of being created and a world that owes its existence to a creator. That difference lies at the root of the invariable stillbirths of science in all ancient cultures."[78] This was ultimately the failure to recognise scientific investigation as a product of free act accruing from a free creator. Thirdly, that ancient civilisation "failed to approach the initial level of intellectual maturity for which strange things are not stumbling blocks but rather aids for deciphering some of the orderliness of nature."[79] When it is strange, then it begs for a deeper answer as to why it isolates itself to qualify to be strange. Science is born by that next step to dig deep in to the explanation of what is not ordinary and to look at it with precisions. Fourthly, there was fear of the eternal return with an

77 Id., *Bible and Science*, Christendom press, Front Royal (VA) 1996, 139.
78 *Ibidem.*, p. 107.
79 Id., *Science and Creation,* 58.

imminent end that made a venture into long scientific research unsustainable. Jaki goes ahead to defend his position that science could only be viably born in the Christian West giving reason why this was so. Let us examine some of his arguments.

Jaki argues first of all that it is a matter of fact the great scientist who promoted scientific revolution that we have today were all from the West. Personalities like, Copernicus, Kepler, Galileo, Descartes, Huygens and Newton are historically inseparable from the development of science in the West. If we are then to ask whether modern science is Western, we would not fail to respond in the affirmative based on the fact these great scientists were westerners. It should be recognised that this West was also Christian West and that Christianity played a vital role in the viable birth of science. Whether science is Western or not is not my interest in this exposition and neither am I interested in arguing that without the West there would be no science. It is an undeniable fact, however, that the West played a great role in the viable birth of science.

Jaki argues that motivated by their Christian faith, the scientists in the medieval epoch sought the harmony of knowledge that prompted them to investigate nature more fully and thus promoting empirical science. According to Jaki, this attitude made the Christians open to natural theology with the implication of its constant search for knowledge with the ability to transcend natural phenomenon. Unlike the ancient cultures that were locked within natural phenomena, the medieval Christian era sought to dominate nature and to transcend it as well. In his own words, Jaki writes, "The transition was made in the perspective which was germane to natural theology and which was instinctively adopted by Newton, chiefly responsible for completing the transition."[80] The transition here is to the viable birth of science which moved away from being dominated by nature to understanding and dominating nature. For science to be born, there was need to trace without prejudices the radical path of Christianity for which the world is a product of a free will of a rational creator and therefore knowable. The Christian doctrine therefore looked at the cosmos as an object deprived of any divinity of its own and open to intellectual

80 Id., *The Road of Science and the Ways to God,* 160.

investigation. Jaki asserts, "Science was born due to the fact that Christians refused to accept the ancient pagan dogma of the divinity of the heavenly regions and bodies, a view technically called pantheism [...] that the universe is not God but only the fully consistent artefact of a rational creator."[81] The Christian scientists investigate into nature in order to conquer it motivated by the conviction that God commanded him to do so. In the words of Jaki: "The same doctrine [Christian doctrine] also gave people confidence that as God-appointed stewards of nature, they had to subdue, control, and investigate it."[82] The last part of the command logically alludes to the rational quest for the universe.

The quest for the intelligibility of the universe was firmly grounded to transcend man and nature. There are two implications to this reality, if God transcends man and nature; the logical consequence is that there is nothing that can limit God to nature and to the mind of man. Therefore, nature cannot be investigated a priori as the Greeks always tried to do. Secondly, God should then be outside nature in order for him to be truly God. If God is outside nature and therefore not limited to it and by it, and not limited to human mind as its product, then a better understanding or rather a just understanding of Him requires a higher faculty of man that stems from reality but which is not limited to it and by it. In fact, according Jaki, "led by the Christian view of a sharp difference between Creator and creatures, Philoponus asserted that for the purposes of science, too, the whole cosmos, created in all its parts by God, and therefore essentially different from him, was rather to be viewed as composed of the same type of matter and governed by the same laws."[83] Being matter and not randomly in flux but with laws that govern it, the late medieval scientists had within his reach all the reasons to experiment with nature to decipher its content ushering science to a viable birth. Most importantly is the fact that the recognition of the laws of nature made him to form scientific generalisations that underlie its operation.

Consequently, the failure to understand a God beyond nature would affect a proper understanding of the cosmos taken as an organism or as

81 Id., *The Origin of Science and the Science of its Origin*, 69.
82 Id., *Patterns or Particles*, 81.
83 Id., *The Relevance of Physics*, 416.

perfect and incorruptible and this ultimately affected the understanding of science as an investigation into external realities of the cosmos in the ancient civilisations discussed above. The conviction of a personal, rational and free creator was only made manifest in the Christian Middle Ages that by her doctrine understood the contingency of the cosmos.[84] The proofs of this God were discussed by Peter Lombard, Augustine, Thomas Aquinas, Anselm and Bonaventura to mention just a few in the course of history.

According to Jaki, it was only in the Christian epoch that contingency of the world could be realised and thus the breaking away from pantheism.[85] He argues: "The contingency of the world derives of course, from its createdness, a point most natural for a Christian to entertain and a point most naturally espoused during these Middle Ages which saw Christianity become for the first time a broadly shared cultural matrix."[86] The understanding of the cosmos in the ancient civilisations was made up of two concepts: the physical part that are eternal and those that are corruptible. While the Christian revolution considered the entire cosmos as contingent and moving to a time of its total destruction owing to the will and the design of its author. Medieval scholars have demonstrated that if the world is contingent, then it must rationally have a source that is not contingent. Whatever is contingent can therefore be subject to scientific investigations. In fact, Jaki argues that this is logical because the world is rational, owing to the fact that it can be comprehended by the human mind only as a product of a rational creator and it cannot be derived from the human mind because he is a creature.[87] We do not then impose on nature what we know about it, we investigate into nature in order to know about it. Paul Haffner argues that it was Thomistic balance alone that made possible the view, so important for the future of science, in the words of Jaki that "the contingency of the universe obviates an a priori discourse about it, while its rationality makes it accessible to the mind though only in an a posteriori manner."[88] Thomas Aquinas established therefore a strong ground on the

---

84 Cf. Id., *The Road to Science and the Ways to God*, 34.
85 Cf. Id., *Scientist and Catholic: Pierre Duhem*, 21.
86 *Ibidem.*, p. 11.
87 Cf. *Ibidem.*, p. 21.
88 P. Haffner, *Creation and Scientific Creativity*, 65. Cf. S. L. Jaki, *The Road of Science and the Ways to God*, 42.

rationality of the universe in a Christian context, that made it a necessity for the development of science not by speculations but by experimentation. In astronomy, this was a major breakthrough as the movements of the planets were not left to popular beliefs and mythology but experimentally proven and demonstrated.

Jaki argues that the medieval time had also Christian scholars who could have led to the stillbirth of science but did not gain strong grounds, because of their Christian doctrines that drove them back to the clear understanding of the separation between creator and creature. Christianity according to Jaki did not go for a simple compromise, because the faith in a personal, rational and free creator could not allow simplistic solutions to pertinent issues. Jaki discusses a series of historians and scientists who constantly enslaved themselves in ignoring Christianity to the extent of manifesting some irrationality. "Such was rationalist historian flights from historical reality to the land of fiction, as distant from history as from rationality."[89] This frustrated the development of science because it moved from one extreme to another. Science cannot but must all the time have as its point of departure concrete reality. Scientific theories are not manufactured in the brain but from real experiments to generalisations in the human intellect.

The role of the historian of science according to Jaki is not to explain the act of discovery but to identify the sources of great intellectual changes that fundamentally affect the sequences of discoveries leading to what he calls the viable birth of science, which occurred only in the late Christian medieval time.[90] It is important to note that even in history there is the need for intellectual rigour to reconstruct moments of fundamental intellectual leap. One wonders why science could not be viably born among the Greeks, even when there were some scientific practices and philosophy. For Jaki, it was the lack of the light of the gospel that gave a completely new worldview with man at the centre of it as one to subdue the world. Christianity gave the scientist the liberty to discover the world because it is not God. We acknowledge though that the zeal to discover the world has time and again gone overboard into scientific adventure that does not

89 S. L . Jaki, *The Origin of Science and the Science of its Origin,* 46.
90 Cf. *Ibidem.*, pp. 62–63.

sometimes serve the common good. This is evidenced by the constant invitation to ecological concerns that can constitute another area for research on the effects of scientific advancement. It is enough to say here, that the command to subdue the world does not exclude responsibility for creation but more ardently implies and advocates for it. There is therefore, no biblical justification of the uncontrolled exploitation of natural resources, not to mention the manipulation of nature for selfish ends. Man ought to be morally aware that destabilising one element of creation is the same as destabilizing the whole of it.

Jaki further asserts: "While biblical monotheism owed nothing to Greek science, that science could develop into true science only within a monotheistic matrix, which happened to be biblical through the mediation of Christianity."[91] One can say that Islam and Judaism were all monotheistic religions and yet Jaki argues that science could not be viably born in them either. This is because Jaki brings in the Christian mediation that brought in the Christological concept of the cosmos. The cosmos that comes from the eternal Logos and returns to it.[92] For Jaki this was a conceptual breakthrough because for the Greeks, the cosmos was eternally begotten while for Christians, Christ was eternally begotten – *monogenes* and the cosmos was created in time.[93] The cosmos in the Christian sense is therefore finite and not eternal; opening it to scientific investigations.

It was that conviction inspired by the Gospel that made possible that science in which data wholly devoid of purpose are investigated for a purpose that can only be recognised through a vision in which man alone is able to achieve.[94] This for Jaki is why science could only be viably born of Christianity, because it is only within the gospel's context of love that science shall fully develop at the service of purpose. "Within that perspective, science too recovers its progressive unity and secures its hope of beneficial progress."[95] This is a fundamental call to the use of science at the service and the benefit of man and this can only be realistic as a means in

91 Id., *The Road of Science and the Ways to God,* 153.
92 Cf. Id., The Saviour of Science, 79–87.
93 Cf. *Ibidem.,* p. 80.
94 Cf. Id., *The Origin of Science and the Science of its Origin,* 106.
95 *Ibidem.*

the participation in God's creative work. Science cannot therefore blindly develop without a reasonable reason that serves man. It was however, that strong reason of purpose of the medieval Christian epoch that motivated scientists to constantly explore into the understanding of nature that constituted the viable birth of science. The purpose of science in this epoch was to make human life better and more fulfilling. This desire was the motivation for the scientists. As they solved one human problem, they were motivated to solve others that either remained a problem or were created by them in the process of solving one and thus leading science into a viable birth.

The Christian understanding of time as linear cannot be ignored as a great impetus for the viable birth and development of science.[96] This is because the linear concept of time is related to the Christian doctrine of creation from nothing by God. God is therefore, the author of time whose fulfilment is in Him. The creation out of nothing implied a beginning, and since there is a beginning, there must be an end, as opposed to the cyclic Aristotelian cosmos that has no beginning nor end. This was only possible in the Christian setting because of the understanding that the universe is not eternal. Commenting on the doctrine of creation in the thoughts of Jaki, Haffner writes: "On the other hand, science achieved a viable birth in a cultural matrix which had already been steeped for centuries in that very Christian doctrine. In other words, the philosophical consequences of this belief were able to penetrate the mentality of the scientist, so as to become almost second nature."[97] The Christian scientist was, therefore, free to investigate into nature as a given. The bottom line of the argument is that science could only be born from the fundamental break from the concept of the world as being part of God. The cosmos is therefore an object of investigation subject to human reason and intellect. Jaki argues that "the cyclic vision of the world could not give birth to a viable science because it numbed the will to discoveries."[98] In other words, it gave no motivation for scientific investigations because all will return and begin again.

Secondly, the linear concept of time, according to Jaki, was a replacement of the Aristotelian laws of motion with the law of inertial motion as

96 Cf. A. Giostra, *Stanley Jaki: Science and Faith in a Realist Perspective*, 36.
97 P. Haffner, *Creation and Scientific Creativity*, 68.
98 S. L. Jaki, *The Origin of Science and the Science of its Origin*, 39.

conceived by John Buridan and Nicole Oresme, which consequently helped Copernicus and early Copernicans to open a new understanding of the cosmos and thus moving science to the next level taken up by Galileo. Jaki underscores the contribution of Buridan because it is one singular achievement of the late medieval period that led to breakage from Aristotelian worldview of motion to Galilean and Newtonian motion that revolutionised science to the level we have now. He argues that "Buridan introduced the application of impetus theory to celestial motion with a reference to a long jump, a most ordinary terrestrial motion. To speak in the same breath of the two motions presupposed the conviction that all bodies in the universe were at the same level."[99] This is fundamental because it is a major breakthrough from the Aristotelian motion that was divided between celestial perfect motion in circle and terrestrial imperfect linear motion. In addition to motion is the breakthrough from the Aristotelian understanding that the moved body should remain in contact with the first mover in order to sustain it in motion. Jaki asserts that "for Aristotle, everything had to be in contact with everything else, with no real beginning anywhere."[100] It was also a revolution from the eternally created divine cosmos to a corruptible contingent cosmos. It is therefore arguable that if science gained more impetus in Newton, it was because the medieval scientists had given it a foundation. Buridan argument was that God gave a push to the planets and because they move in space without friction, they are kept in constant motion without having to be in contact with the first mover. The first mover was therefore God, who logically would not be within creation itself in order to be able to move it. Jaki would conclude that "belief in a creator whose powers were not limited to the features of the actually observed world contributed, therefore, most effectively to the liberation of critical thinking from the shackles of Aristotelian science."[101] This led to the discovery of the law of gravity and electromagnetics that gave birth to modern science because Buridan broke away from Aristotelian science evident by his discussion of projectile motion and of the fall of bodies.[102] Consequently,

---

99 Id., *The Saviour of Science*, 75.
100 Id., *Particles or Principles*, 78.
101 Id., *Science and Creation*, 234.
102 Cf. *Ibidem*.

"further refinement in the concept of impetus led to the correct definition of momentum together with initial motion on which rest the whole edifice of physics."[103]

Jaki insists that it is an historical mistake not to present Copernicus, Kepler and Galileo first of all as Christians and merely as followers of Plato, because from Grosseteste to Galileo science can only be explained from a Christian understanding of the created world.[104] Kepler is quoted to have exclaimed after his revolutionary discovery of the laws of planetary motion: "I thank thee, Lord God our creator, that thou hast allowed me to see the beauty in thy work of creation."[105] Such a conviction that makes a scientist acknowledge the beauty in creation, alluding to the transcendental of being as good can only be attributed to his faith in God who is not only creator but also the giver of knowledge. For Jaki, the men of the renaissance cannot burry the medieval age into darkness when one of the texts frequently quoted was the Old Testament book of *Wisdom* 11:20: "God disposed everything in measure, number and weight." This is a text that tells much of what science does, as defined by Jaki. Christians believe that the scriptures were inspired by the Holy Spirit and if the book of wisdom, without any evidence of experiments already declared that the it was God who made all things in their quantitative form, and we know that science is the 'quantitative measurement of things in motion,' we cannot but have to admit from such an assertion that Christianity had an heavy hand in the viable birth of science.

It was Galileo who took very seriously the issue of numbers and held mathematics as very close to the knowledge of the creator. For him "man's understanding of quantities is no less perfect than the creator's knowledge of them."[106] Even if this tilted the radar of thinking towards a priori knowledge, it does not deny him the credit of being influenced by his Christian faith. It is unfortunate, according to Jaki, that Descartes reduced rationality to human ability and lost the medieval concept of the divine rationality for

103 *Ibidem.*, p. 242.
104 Cf. Id., *The Road of Science and the Ways to God,* 47.
105 S. M. Barr, *The Believing Scientists*, 180.
106 S. L. Jaki, *The Saviour of Science*, 98.

which human reason is just a participation. This was a sure path into empty science deprived of reasonableness.

According to Jaki, the Medieval scientists anticipated four very important scientific concepts that led to the viable birth of science:

> The concepts of inertia and momentum, the assertion of the possibility of three-dimensional infinite vacuum, the groping for quantitative in addition to qualitative account of physical processes and finally the realization of a basic need for experimentation if progress was to be made in understanding and conquering nature.[107]

These were indispensable scientific concepts that led to the development of science in the modern Christian West that is most times obscured by the view that the medieval epoch before Galileo was a dark age. It was Duhem's preoccupation to demonstrate that this stereotype is not historically accurate.

Jaki points out critical moments in history that made the birth of science only possible in the Western Christian Civilisation, citing some very revolutionary events such as the understanding of gravitational force and the heliocentric concept of the universe.[108] These events came at the right time to take science to the next level of its viability that could not occur elsewhere except in the Christian Civilisation moved by the Christian doctrine.

These great scientists went the long length on the road to science, because they were supported by their Christian faith and were constantly investigating into nature motivated by their faith. Decisively, Jaki argues that Christianity saved science by holding at all times that "natures bespeak an intelligibility that derives from a transcendental source."[109] This logically follows that science, which is an investigation into nature, can have a fruitful dialogue with faith which is a leap into the transcendental realm.

The fact that universities and schools were directly under the guidance of the Christian church is another strong reason to assert that science could only be viably born in the Christian west. Jaki asserts:

> It was the fruit of a commonly shared belief nurtured in part by uniform educational system, consisting of universities, cathedral schools, and monasteries, the

107 Id., *Science and Creation*, 233.
108 Cf. Id., *The Road of Science and the Ways to God*, 97–103.
109 *Ibidem.*, p. 48.

> like of which neither Greece, nor Rome, nor any ancient great culture was ever able to produce. [...] in the medieval universities, the readiness to innovate was far a more widespread phenomenon than historiography steeped in the intellectual militancy of the Enlightenment would have ever wanted to consider.[110]

Science definitely needed an environment in which creativity was allowed and promoted and the medieval universities and schools, under the strict governance of the church were there to promote research and the development of science to the scientific revolution that we now enjoy. We admit that this was a slow but a definitive process on the right path that goes a long way to justify the fact that science, if not born of a Christian mentality, then at least it was born within the Christian institutions including convents and monasteries located geographically in Europe.

The universities and cathedral schools, apart from acting as reliable centres of learning, served as conservatories of the great achievements of the medieval times. Jaki affirms that the commentaries of Buridan and Oresme were copied and made available in the major universities in Europe.[111] It is arguable that these resources were used by modern scientist to build a new vision that saw the scientific revolution in the European West.

When Jaki talks of a scientific origin, he does not intend to underscore a single scientific discovery but seems to go more deeply into science as a complete system so as to understand it as a body of specific quantitative knowledge. This could only be developed, promoted and preserved in a university setting that was provided by the Medieval epoch. He points out how scholars like Francis Bacon missed the fundamental link of the birth of science because of prejudices that blinded them. "Bacon and others proud of the reformation went too far in depicting the ten long medieval centuries as deprived of light, spiritual and philosophical which would have saved them from looking upon themselves as the lucky accident of times, a view hardly philosophical or spiritual."[112] Bacon looked at the scientific method of his time without minding to trace its origin in the Medieval era because he was already bias about it as a dark age.

110 Id., *Science and creation*, 232.
111 Id., *Patterns or Principles*, 76.
112 Id., *The Origin of Science and the Science of its Origin*, 11.

This kind of contradiction and inconsistency is what is referred to as irrationality that contradicts the nature of man and the good use of reason that amounts to intellectual dishonesty. Truth cannot and should not be selectively accepted. To declare a period of time as an intellectual vacuum is to attribute to the human person an interval in which man ceases to be a rational being and consequently not human, only to regain it later in history. History cannot account for such a time that the human person was incapable of reason by nature or even by choice but it is possible that some moments of history are categorised by a shift in concern of the human thoughts and philosophy. This is typical of the medieval time that man's thought was a unified corpus.

1.3.3 The Role of Catholic Church in the Development of Science

Jaki's major preoccupation in doing the history of science was to demonstrate that the Catholic Church has never been against the development of science, as it is often argued by many historians of science. Antonio Colombo writes that way back in 1957, after his PhD in Physics, Jaki had already seen a great problem in reconstructing the history of science to respond to a better understanding of the relationship between science and religion. He writes:

> Father Jaki had at that time developed the idea that the problem of the relation between science and religion had much to do with the historical aspects of matter. Proximity of Stanford University, whose library is well stocked allowed him to probe many aspects of the history of science, a subject to which real scientists paid little or no attention.[113]

To understand the development of science over time, Jaki saw a need for an historical reconstruction. This commitment to the tracing of facts about the history of science helped him to come out with some arguments to show that the Catholic Church and therefore Christianity contributed to a great degree to the birth of science and its development to the current modern state. The singular argument raised against the Catholic Church as an obstacle to the development of science revolve time and again around

113 A. Colombo, "Introduction", in A. Giostra, *Stanley Jaki: Science and Faith in a Realist Perspective*, IF Press, Rome 2019, 32.

the condemnation of Galileo Galilei of the heliocentrism in 1633. This however must be studied critically within the context of Galileo's time.[114] A detailed analysis of Galileo's controversy is outside the scope of this work, but could constitute an area of deep research for clarity on the Galileo's affairs. Stephen M. Barr raises some critical issues regarding the attitude of the Catholic Church towards science and why the singular condemnation of Galileo cannot typify the attitude of the church because of the historical setting and the scientific lack of evidences at that time. Galileo himself did not have a scientific proof even when he was proven right in the course of time.[115] It can be argued that if the Catholic Church was hostile to science, many Catholic clergies, monks and religious could have not ventured into science with enormous contributions to its development, let alone faithful lay Christians whose contributions to science should be taken seriously. The Benedictine monks for example, with their motto *ora et labora,* made many discoveries on the line of medicine but also conserved a lot of European scientific literature from destructions.[116] That science flourished within the walls of the monasteries, is a clear demonstration of the church's commitment to its course.

Jaki boldly argues that "if therefore that science which we are rightly proud of, could be born, it was because the Catholic Church served as a midwife."[117] One who is a midwife knows how sometimes the midwife has to be harsh in order for the expectant mother to bear a child safely without suffocating it for a stillbirth. The hostility between science and faith is a fact that cannot be hidden and yet we must ask ourselves why scientists have this hostility. Jaki points out this uneasiness: "Nothing irks the secular world so much as a hint, let alone a scholarly demonstration,

114 For some discussions about Galileo's, cf. S. L. Jaki, *The Case for Galileo Rehabilitation*, in "Fidelity," March, 1986, 37–41. Id., *The Case for Galileo's Rehabilitation*, in P. Larrey, "Science and Religion: An Anthology, Selection of Texts from Aquinas, Bacon, Galileo, Darwin and John Paul II," ARCANE (ed), Rome, 2006, 254–263.

115 Cf. S. M. Barr, *Modern Physics and Ancient Faith,* University of Notre Dame Press, Notre Dame (IN) 2013, 1–10.

116 Cf. F. Agnoli, *Scienzati, Dunque Credenti; Come la Bibbia e la Chiesa hanno creato la Scienza sperimentale,* Cantagalli S.r.l., Siena, 2020.

117 S. M. Barr, *Modern Physics and Ancient Faith,* 239.

that supernatural revelation as registered in the Bible is germane to science. It is not only germane to science but it also made possible the only viable birth of science."[118] This was motivated by anti-religious attitude towards the explanation of natural phenomena.

The crisis of modern science, according to Jaki, stems from the negation of its origin by the scientists themselves or at least by pseudo scientists and historians of science or what Barr would call materialists.[119] A materialist is one who believes that beyond matter there is no other existence, matter is everything and non-material operations of the mind is a mere product of matter and therefore, metaphysics is fantasy and pieces of nonsense of the fearful man of material decay. The consequence of this is the assumption that the scientist has infallibility to every answer and that every aspect of life can be explained in a materialistic and mechanistic way, yet these are just two of the many ways of looking at the same reality. Reality in nature has many faces that demands more than one method and way of interrogating it. The stereotype that science has all answers is so deeply entrenched even in the education system that in an inquisitive class of average students, more questions are asked in the class of religion than in a science class. While the science class concentrates on memorizing formulae and theories, the class of religion is preoccupied with doubts, subjective objections and deconstructions provoked by unscrutinised assumptions of science. Jaki's arguments about the Christian foundation and the role of the Catholic Church in the development of science has met some hostilities that I would like to examine in the subsequent section.

1.3.4 Hostility to the Christian Foundation of Science

The assertion that science was born in the Christian ambience of the late medieval time is a controversial matter that Jaki defends not without many criticisms from other thinkers because of the different approaches to the interpretation of history or even different interpretation of the term 'science' and above all what viable birth would mean to various scholars. It is therefore important to keenly follow Jaki's historical arguments that

---

118 S. L. Jaki, *The Limits of a Limitless Science and Other Essays,* 48.
119 Cf. S. M. Barr, *Modern Physics and Ancient Faith,* 1–29.

collocates the birth of science in the Christian setting decisively because of the Christian doctrine that gives a revolutionary view of the world and all that it contains. He argues about some of the attitudes that already foretells some of the criticism thus: "The message, first formulated by the gurus of the enlightenment consists in the claim that science is the only reliable saviour of mankind and that for science to be, Christianity or religion most explicitly steeped in belief in a most extraordinary saviour, first must be discredited."[120] The discrediting of religion is therefore, a strategy of the anti-religionists that deliberately puts into bracket reason where and when it is most needed.

It is apparent that the hostility of enlightenment is not anymore out rightly manifest but this does not mean that the core intention is gone, the struggle has turned into a cold war with scientists seeking the collapse of faith in God by their ever-progressive discoveries. This stems from the misconception that faith has to die before science can triumph. That science that was born because of the strong faith that material existence is a given by God for investigation, now wants to claim dominion over its origin. It is like a child living upstairs and declares that he can live without the basement of the house. This is what concerned Jaki almost his entire life that science and faith are not at war and should not be at war. According to Jaki it is mediocrity on both sides of the scientists and the theologians that create what is called today conflict between faith and science. Many scientists go beyond their limits and try to surround their science with some kind of philosophy and many uninformed theologians in their zeal for progress either "baptise" some scientific discoveries or completely deny the unlimited progress of science. At the crossroad of disciplines, there must be the honesty to stop and to dialogue in order to distinguish what belongs to fields other than my own. Realising that we are a crossroad is not a given, it needs through formation of the scientists and the theologian to know and accept their limits. Many of the situations of conflicts come out of what happens at the meeting point of science and faith.

Jaki, while writing on the influence of Pierre Duhem, taking him principally as a faithful Catholic asserts: "The ultimate significance of Buridan's

---

120 S. L. Jaki, *Scientist and Catholic: Pierre Duhem*, 8.

statement was nothing less than that science could not have its beginning except as fruit of the most fundamental of all beginnings. The first motion in God."[121] This is in reference to the impetus force that was developed by Newton into the law of inertia.[122] According to Jaki, "it was in fact in the late Victorian times that the western world turned its back on its Christian origins, which is the creation of all in the beginning."[123] The Victorian era was characterised by economic prosperity and it was at this time in history that intuitionism and utilitarianism prospered as opposed to Christian moral standards. Man in his prosperity tends to cancel his past and his origin until he is stroked by real existential questions.

According to Jaki, it is by logical and honest arguments that we can be reasonable in defence of the viable birth of science in the Catholic Church. However, something more is required, that is, the ability to transcend what is merely factual to what is metaphysically meaningful. There is a necessary link between the mind's ability to go beyond itself and what it physically encounters. It is only such minds that have the eye to see beyond the microscope a reality bigger than the tinniest microorganism, faith is therefore not the failure to see but the ability to transcend.[124] As already noted, the medieval period saw the Catholic Church as institution in charge of education and therefore championing both theological and philosophical studies. Science was broadly considered as a branch of philosophy making the Catholic Church institutionally responsible for its birth and development.

Jaki was greatly influenced by Duhem who, as far back as in 1936, had noted that the threats of science, used by flawed philosophy and their promoters, "amounts to denying, in the name of science as such, to all religion the very right to exists."[125] They pretend to have established that no sensible man can admit, at the same time, the value of science and also belief in the dogmas of religion.[126] This plays on the emotion of scholars so much so that to pose as a scientists tantamount to having an important

121 *Ibidem.*, p. 85.
122 Cf. A. Giostra, *Stanley Jaki: Science and Faith in a Realist Perspective*, 49.
123 S. L. Jaki, *Scientist and Catholic: Pierre Duhem*, 87.
124 Cf. *Ibidem.*, pp. 58–83.
125 *Ibidem.*, p. 236.
126 Cf. *Ibidem.*

status in the academic field and consequently declaring oneself antifaith. This however is the failure to integrate in the same human person science and religion that are interrogated by the same reason capable of understanding both as separate realities. This is the ideology that has defused from the Vienna Circle in the attempt of creating a world full of proofs without truth.[127] At the peak of this argument is the ideology that any knowledge that cannot be proven experimentally is not worth holding and, since the Catholic Church teaches a religion that cannot be subjected to experimental scrutiny, it should have no authority on scientific matters. This kind of approach defeats rationality as the nature of the human person and the mode by which science progresses because knowledge and reality cannot be reduced to science alone.

The strength of science is that it is the author of its results and therefore in charge of all the processes that are involved in creating a belief system. We are therefore dealing with an artefact that carries with it the limits of its author; owing to the fact that man is fallible, we can assert that scientific findings carry with them their fallibilities notwithstanding the fact that they are always provisional answers to the pressing questions of the time based on immense assumptions. Scientism, a form of reductionism that science is the only objective determinant of both normative and epistemic value of all realities, is a modern disease and is actually not different from idolatry, because if man turns the artefact of his hands to a saviour, it goes without say that he falls into idolatry.[128] The contingent entity cannot claim incontinency for its own product. The potentiality of the human mind goes beyond physical sciences which is only one of its capabilities. This is fundamentally essential for the relationship between science and faith. Jaki writes talking about methods and what the investigator sees:

> He can recognize at the same time what gives variety to those procedures and what unites them. Everywhere he sees the one and the same essential tools in order to arrive at the truth. But in each domain, he sees that reason adopt the tools to the special object whose knowledge it wants to acquire.[129]

---

127 Id., *The Relevance of Physics*, 401.

128 Id., *Scientist and Catholic: Pierre Duhem*, 236.

129 *Ibidem.*

This argument brings out the unifying role of reason as the most fundamental tool by which we can do both science and transcend to God. Science, being a reasonable enterprise, prompts man to produce more accurate tools to investigate into nature, and doing so the investigator cannot but also hold his faith at a level that is not limited to what his science produces. He learns to distinguish between what his tools can do and what he is actually capable of doing with the products of his discoveries.

In order for human mind to operate without hindrance, it must purify itself from prejudices and taking a rigid stand in the search of truth. Besides, the investigating subject must hold on to intellectual honesty in order to apply reason towards the object of its search. "He then recognizes that in order to reach religious truth, the human reason uses no other tools than the ones it uses to attain other truths, but it uses them in a different manner because the principle from which it starts and the conclusions to which it tends are different."[130] When we talk about principles in philosophy, we refer precisely to those universal aspects of reality that are not limited merely to segments of knowledge. This is what philosophers deal with. Modern science taken seriously operates by investigating deeply into a small area of experience and most of the time; this is prompted by some problems that demand for answer to the question, "how?" The conclusions to such problems follow directly from the problems that provoked them. Apparently, there is no doubt that if we have to meaningfully talk about the world, we must always do so in a scientific, metaphysical, religious and philosophical languages. There seems to be no other language to speak about the cosmos apart from these unless the human mind returns to the prescientific age of mythology and superstitions.

1.4 Stanley Jaki and Other Thoughts on Medieval Science

I would like to consider other voices about the birth of science in the medieval time and I will consider here authors who responded directly to the views of Jaki. I will honestly bring out their views and show how they relate or differ with that of Jaki.

130 *Ibidem.*, p. 238.

Edward Grant, *The foundation of Modern Sciences in the Middle Ages* (1996), argues that modern science would not be born without the events that constitute the years 1200 to the 1600 in the European West.[131] This goes a long way in supporting Jaki's view that the European West at that time was purely Christian and thus sometimes called Christian Civilisation. The argument is that Christian doctrine made science a possibility since it [science] depended very much on culture according to Grant. Therefore, Jaki's argument that the Christian culture dominated this period cannot be ignored if we are to be historically accurate. It is therefore, a historical fact that in this period of the human history in the West, Christianity was dominant and science flourished in it. The reason for science getting a revolutionary status at this time in his history and in the European west remains a matter open to discussion and interpretation.

David C. Lindberg in "Science and the Medieval Church" (2013), tries to give a balanced approach to the relationship between the Christian church and modern science. He considers Jaki's arguments as one of the extremists' approach to the role of Christianity in the birth of science. Jaki probably escapes the grip of Lindberg's claws, as far as being an extremists and reductionist is concerned, by specifically defining science as exact science, which is actually limited to physics. Lindberg points out two extremes: the powerful anti-religious currents promoted by enlightenment referred to popularly as "the warfare thesis" and the antithetical claim that the Christian church was not the adversary of science but its partner and patron.[132] The latter is the view that Lindberg attribute to Jaki, to which he vehemently disagrees, especially, when Jaki sustains that it led to the viable birth of science. The underlying contention is probably rooted in the church's partnership with religion going back to the Galileo saga has it has already been pointed out. Can the Catholic Church still remain a partner and patron of science with the condemnation of Galileo? I will continue to discuss this dark spot, time

---

131 Cf. E. Grant, *The Foundation of Modern Sciences in the Middle Ages: Their Religious, Institutional and Intellectual Contexts*, Cambridge University Press, Cambridge 1996.

132 Cf. D.C. Lindberg "Science and the Medieval Church," in Id., – M.H., Shank (eds.), *The Cambridge History of Science,* vol. II, Cambridge University Press, New York, 2013, 268.

and again insisting that to understand the Galileo issue, we must situate ourselves in Galileo's time and mentality. We can interpret history with modern lenses but without forgetting that our lenses are modern and new.

The medieval Christian scholars were both theologians and scientists who sought harmony in their studies. They were driven by their Christian culture in the search of both scientific facts and religious truth. It must also be born in mind that the medieval times had its own politics and authority that controlled learning without necessarily advocating for a uniform ideology. Lindberg argues: "Religion was not a bothersome meddler but an inevitable and (in medieval terms) legitimate players in the game."[133]

The game was however played with care that harmony is kept and better and accurate solutions were sought. It is arguable that the medieval time stimulated scientific investigations in the search of stability and harmony. The medieval Christians were able to coexist with Aristotelian world view even when some of his doctrine were utterly a contradiction to the Christian faith and doctrine. This was the kind of atmosphere that accepted allies without necessarily agreeing totally. Lindberg argues:

> Even the Aristotelian corpus, despite its manifest dangers, provided a satisfying account of so many aspects of the created world, as well as tools of unprecedented power for the further explanation and analysis of that world, that scholars, including those of undoubted orthodoxy, were loath to relinquish it.[134]

Jaki, however, would not agree to this point as he dismisses Aristotelian world view as completely misleading and constituted no science at all. It will demand the intervention of Richard Connell that "philosophy and science were just one thing in the ancient Greece and even of late natural science was referred to as experimental philosophy. Sir Isaac Newton's treatise on physics, coming after Galileo is entitled, *The Mathematical Principles of Natural Philosophy.*"[135]

Despite the banning of Etienne Temper in 1270 and 1277 of Aristotelian propositions, the consequences were more positive than negative according to Lindberg. He argues that it actually stimulated the search for alternatives

133 *Ibidem.*, p. 271.
134 *Ibidem.*, p. 276.
135 R. J. CONNELL, *The Empirical Intelligence – the Human Empirical Mode,* The Edwin Mellen Press, Lewiston, New York, 1988, 170.

while the church in the hands of Thomas Aquinas (1225–1274) and Albert the Great (1200–1280) became major proponents of Aristotelian philosophy purging it of error. Lindberg concludes: "The Catholic Church became the most powerful and generous patron of serious philosophical and scientific coffer the West had ever seen."[136] It is also arguable that Buridan formulated the theory of impetus following the 1277 condemnation as a search for an alternative to the Aristotelian notion of the eternity of the universe. Jaki will argue at length that Newton's law of motion was already implied in the Medieval times.[137] In the same wake about the condemnation of the Aristotelian doctrine by Temper, Jaki seems to think in the same way as Lindberg with a slightly different approach, that the condemnation gave science the beginning of some autonomy. He argues thus:

> It was also an implicit affirmation of the view that the laws of the universe were not necessarily identical with inferences based on a priori philosophical speculations. The decision was a momentous indeed, encouraging as it did efforts that aimed at both keeping scientific research within its own range and securing to it a proper measure of autonomy.[138]

For Jaki, the condemnation of the Aristotelian propositions "shaped the state of mind and conceptual groundwork for a revolutionary new approach towards the understanding of the workings of celestials and terrestrial bodies."[139] May be the scientists of the time would not have ventured into a new search if those propositions were not officially condemned. Jaki, aware that Herbert Butterfield had endeavoured before him to trace the origins of modern science argues that he [Butterfield] failed to demonstrate that the 1277 condemnation was basically because of the Christian conviction that linked creation to a creator God who does his work in full freedom.[140] Jaki also criticises Butterfield for not explaining clearly the

---

136 D.C. Lindberg "Science and the Medieval Church," 278. Cf. Id., *The Beginning of Western Science. The European Scientific Tradition in Philosophical and Institutional Context BC-AD 1450,* University of Chicago Press, Chicago (IL) 1992, 234–241.

137 Cf. S. L. Jaki, *Saviour of Science*, 72.

138 Id., *The Relevance of Physics,* 418.

139 Id., *Science and Creation,* 231.

140 Cf. Id., *Bible and Science*, 132. Cf. H. Butterfield, *The Origins of Modern Science*, The Free Press, New York, 1957.

great contributions of Pier Duhem as coming from a faith conviction of a member of the Catholic Church. Butterfield was careful in his presentation not to give credence to the controversial line that the Christian faith had a role to play in the thinking of Pier Duhem and other medieval scholars that consequently gave birth to modern science. He however agrees that the medieval scholars and scientists had a great role to play in the birth of modern science.

Looking at the attitude of the medieval time it can be argued that there was no attempt to block the progress of science at this time by the church. The Catholic Church was concerned about embracing Aristotelian work with error and guarded against it. The Catholic Church sought to integrate philosophical and scientific knowledge into her theological doctrine and tolerated incompatible philosophies in order to understand them better. In our modern time, this kind of marriage has led to more confusion, misunderstanding and error. Many theologians have sought to marry theologically scientific discourse with basic theological beliefs without proper competence. It is plausible that it is the work of the philosopher because the philosophical method permits a universal discourse.

Lindberg, therefore, vehemently disagrees with Jaki that Christian theology gave birth to western modern science. He however agrees that Christianity both in culture and theology had some influence both positive and negative to the growth of modern science. He argues favourably that the handmaiden formula of St. Augustine protected science whenever it was threatened by faith and authorities.[141] Lindberg argues further that there was no particular religious motive that directly sought the promotion of science and there seems to be no evidence that scientists at this time worked to answer any religious questions by their scientific investigation.[142] Lindberg concludes that "the medieval scholar who specializes in the natural science certainly did not regard himself as suffering from overly close supervision by theologians or any other party within the church."[143]

The arguments of Lindberg are interesting and clear the ground that the Catholic Church never stood on the path of the development of modern

141 St. Augustine argued that science is the handmaid of the bible and theology.
142 Cf. D. C. Lindberg "Science and the Medieval Church," 284.
143 *Ibidem.*, p. 285.

science but actually gave it room and impetus to grow. Lindberg is careful not to assert that the Catholic Church was a patron or explicitly a partner of science but that it did not block scientific progress. Jaki's argument is however that besides the Church giving space to science in the middle ages, Christian scientists engaged directly into scientific discoveries that took science to a new level from which it flourished and that they did it because their Christian faith did not impede them as it was in other ancient civilisation by giving them a world view that edified their curiosity and research. If we want to appreciate this argument, we must bear in mind that Jaki qualifies the birth of science to have been viable only in the Christian setting. He does not deny the fact that other civilisations gave birth to science but that it was a stillbirth, something similar to post-natal abortion. So, science was born but did not survive the test of time and died in some civilisations while in others it was born dead.

When we have to be just to the birth and development of science in the Christian West, we must consider the fact that Greek and Islamic science was exposed to Christendom through translation much later. This means that science went from the Greeks to the Arabs and to Latinians, where according to Jaki it received a viable birth or was resuscitated.[144]

In *The Saviour of Science* (2000), Jaki takes the same line of thought as that of Lindberg but insists that Christian doctrine was implicit and manifest in the medieval culture.[145] We have argued that the medieval age promoted the development of modern science more than any age in history, but we ought to recall that to talk of the medieval west is almost equal to talking about Christianity and no less than Christian doctrine and faith. In his own words Jaki asserts that "to say medieval is almost to say Christian

144 Cf. C. Burnett, "Translation and Transmission of Greek and Islamic Science to Latin Christendom," in D.C., Lindberg – M.H., Shank (eds.), *The Cambridge History of Science,* vol. II, Cambridge University Press, New York, 2013, 341–363.

145 There has been some academic confrontation between Lindberg and Jaki since 1986. Cf. D.C Lindberg – R. L. Numbers, *God and Nature: Historical Essay in the Encounter Between Christianity and Science,* University of California Press, Berkeley (CA) 1986.

and in a rather dogmatic way."[146] This also explains the anti-Christian hostility to this historical period described as the dark age by some scholars.

Jaki would respond to Lindberg about the Aristotelian philosophy in these words. One brilliant example of how science can subdue reason is not to question its assumption is about the law of motion of Aristotle. The latter has existed for 700 years. It states that two falling bodies with one half the weight will fall at different rates with the heavier falling twice as fast. Jaki observes: "Athens must have provided enough cases of falling tools and stones of every size to show that the time of fall from the same height was much the same for the great majority of objects, whatever their size, shape, or weight."[147] It is noticeable that scientific assumptions are sometimes taken without even seeing glaring contradictions. It is arguably not different in the 21st century that assumptions of science are accepted without question trusting that the scientist knows it all making the scientist have the claim of knowing it all.

There seems to be no doubt that science has over the years been successful in bringing out new understanding of the universe and what it contains but we bear in mind the assumptions and the dogmas of science. Jaki argues that "mechanism had successes to an astounding degree, but at the same time carried along through its three-century lifespan serious questions as to the soundness of its basic assumptions."[148] Every time we meet such profound assertions, we want to underscore the philosophical implications imbedded in them like, "soundness of its basic assumptions." This implicitly refers to the rationality of the many assumptions used by science. Science operates on countless assumptions that must be rational and this is the reason why the argument about the rationality of science is not because scientific measurements or measurements of quantities are simply rational but that these measurements are based on assumptions that must be rational to render science an intellectual activity. When science is driven by reason, then scientific assumptions are realistic and therefore scientific endeavours.

---

146 S. L. Jaki, *The Saviour of Science*, 2.
147 Cf. Id., *The Relevance of Physics*, 32.
148 *Ibidem.*, p. 91.

Stanley Jaki did not live long enough to see the deviation of science in the west when Ethiopian Airlines flight ET302 turned into a bomb of science that in post-mortem style gave the explanation to the disappearance of the Lion Air in Indonesia of the same make. The 737 Max, without the technical details, rendered the pilot helpless before the plane that nosedived 10 meters into the heart of the earth. This is not enough signal for scientifically blind and greedy world that no science can replace human reason and that sound reason must accompany scientific innovations. The plane was a time bomb in the hands of artificial intelligence. Needless to say, that science cannot be the saviour of the world.[149]

In dealing with matters of faith and science as two aspects that involve the same human intellect, we should avoid the mistake made by both scientists and some theologians of making God, the God of the "fuse" to connect where there is disconnection in science, what other scholars call the God of the "gaps" or God of the "fill in", as Jaki would term it. The Christian God created and continues to sustain creation. This is reasonable because our world is not a blind world directed by science whose failure and success are proportionally the same. Shallow philosophy like that of the stoics would not save the world as it plunged it to more irrationality than common sense. "As for the rationality of the stoic universe based on the universal rule of fire, it was merely a brave front put up in the face of fundamental irrationality and hopelessness."[150]

It is no accident that irrationality is qualified as fundamental because the stoics had no standard measure of their arguments and therefore no foundation. The leap of faith was what the stoics never had or comprehended that the universe and all it contained was a result of free creation that science continuously tries to understand. Jaki emphasizes that the createdness of the universe should strictly and emphatically be understood univocally as creation out of nothing.[151]

Only medieval Christian scientists were motivated to understand nature as what is created and given to us by God to conquer and make use of. We cannot, of course, doubt the fact that even the most faithful and intelligent

149 Cf. Id., *The Saviour of Science*, 41–42.
150 *Ibidem.*, p. 75.
151 Cf. *Ibidem.*

Christian at one point asks questions that may affect their faith but does not necessarily make them unbelievers. It is therefore no proof to make collections of snippets of controversial statements made by Christian scientists to brand them unbelievers. To do that is a "transparent device to minimize the support he derived from his Christian faith as he conceived and kept developing the views that opened the way to a truly scientific grasp of the universe, the greatest price for science."[152]

It can be sustained that the non-observable aspect of intellectual operation of the medieval scientists and many other Christian scientists was their faith in God, the Creator of the universe. They had no reason to make a declaration about their faith because science made sense in their unified world view. Jaki argues that, for a Christian, salvation is a mystery but a mystery that does not deprive it of being reasonable.[153] This is the rationality that is required for both science and faith so that the truth of their contents is intelligible to both a theologian and a scientist who can then have fruitful dialogues. As Jaki argues that science was born of Christianity and therefore, within faith context, we want to argue that both are rational activities of man that have a profound relationship in the proper use of reason.

1.5 Recapitulation

Our concern in this chapter has been to present the major arguments of Stanley Jaki for the singular viable birth of Science in the Christian Civilisation of the late medieval period. We endeavoured to establish that, in carrying out this research, we shall be fulfilling the vocation of a philosopher in seeking a fruitful dialogue between science and faith, taking the thinking of Stanley Jaki as a springboard and using philosophical arguments and principles.

We were able to iron out methodological issues and the need for thinkers and scholars to remain within the domains of their competence. We reflected on the need for intellectual honesty that accepts limits of methods and tools and has the audacity to handover the search of knowledge to the competent other. We also noted that honest reason should always guide the

152 *Ibidem.*, p. 85.
153 Cf. *Ibidem.*, p. 89.

dialogue between science and faith because it is reason that seeks balance and does not want to contradict itself. We discussed the fact that to envisage conflict is to put reason into confrontation with itself. This, we noted, it is not an easy task that took Jaki the greater part of his life trying to show the kind of relationship that should be promoted between science and faith demonstrating its pertinent attachment to Christian faith. Science will be an empty endeavour without the proper use of reason. There is, therefore, the obligation to use reason properly. It is not enough to hold a position or carry out an experiment without a proper rationale for which it is done. This calls for a high level of self-evaluation and obedience to the dictates of proper reason.

We dwelt at some length on the concept of science, according to Jaki, as he defined it as the study of the quantitative aspects of things in motion. We noted that he limited science to physics in order to save it from every misinterpretation and distinguish it from other disciplines. This however inclined his thoughts towards some form of reductionism when he denied belonging to science all other disciplines and yet we know that words have several meanings, with science not being an exception especially when, etymologically, science has an epistemological dimension. This goes beyond experimental sciences to a systematic rigorous discourse that experimental science itself needs. We saw Jaki's arguments for the limits of science, that it is limited in as far as it tries to get outside its competence to deal with qualitative aspects of things and unlimited as long as it remains within the investigation of quantitative aspects of things in motion. This is because there is no time that science will give definitive answers to scientific problems. All scientific answers offered by science are provisional; based on many assumptions and generalisations. Science is therefore limited by its methods and tools. To keep the limits and boundaries of science and other disciplines, there is therefore a need for unconditional and non-selective intellectual honesty.

On the origin of science, we saw that according to Jaki science did not have a viable birth in the ancient civilisations like, India, China, Egypt, among the Muslim, Mesopotamia and Greece, despite the fact that they had some elements of scientific technicalities. The basic reason for this was their pantheistic vision of the cosmos that made them get locked within nature itself, the failure to get away from the cyclic movement of time and eternal return that stunted scientific initiatives and curiosities. The ancient civilisations all failed to make conceptual generalisations that were necessary for the viable birth of science. The concept of creation was another impediment to the viable birth of science, because the Aristotelian emanation from the first cause could not distinguish creator from creatures and consequently, the failure to see purpose and reason in creation which could have promoted science to the next level. In many ancient civilisations, natural phenomena were attributed to some spiritual irrational despot making and investigation into cause almost an impossibility.

We then delved into the reasons for which Jaki asserts that science could only be viably born in the late Medieval Christian West. The major breakthrough was the Christian concept of creation of the universe out of nothing, in the beginning and in time by a personal and rational creator. The creator who creates for a purpose and remains outside creation and time. This exposed creation to scientific investigations of causes of natural phenomena within nature itself. It is was a decisive point of breakage from pantheism. Creation in the beginning changed the vision of the cosmos from a cyclic concept of time to a linear movement of time. Major protosciences were nurtured in the medieval epoch with the principal breakthrough of Buridan inertial motion. The Christian West was able to form generalisations using mathematics and experiments because they had the institutions in the name of universities and the cathedral schools to promote them. The driving force behind the scientific research at this time was the desire to transcend nature and to conquer it as a command coming directly from God.

We also discussed the roles of the Catholic Church as an institution in the promotion of science and we saw that the Church has always promoted and supported scientific initiatives despite the accusations meted upon it about the Galileo Galilei case. We asserted that the Galileo case cannot singularly define the attitude of the Catholic Church towards science. We also discussed the fact that some opposition targeted towards the Catholic Church had science only as a scapegoat. The major target has been and is the social teachings of the church on the morals of using scientific technologies and discoveries.

We finally discussed Jaki's thoughts in relation to other thinkers about the viable birth of science in the late medieval Christian West. We noted that it is important to understand well what he means of a viable birth in order to appreciate his arguments. We however noted that it remains debatable whether the medieval scientists and thinkers were motivated particularly by their faith because we can only judge them from outside without knowing their inner dispositions. However, the fact that the medieval Christian West provided a conducive ambient for the flourishing of science cannot be denied. The defence that science was born in the late medieval Christian West according to Jaki goes directly to the "defence of the primary mental outlook over the bare facts of sensory evidences."[154]

154 Id., *Science and Creation*, 238.

Chapter Two Relationship between Science and Religion

2.1 Introduction

Stanley Jaki spent a lot of time demonstrating the relationship between science and religion. He demonstrated that the various attempts made in discussing this relationship have been faulty because either the people doing it did not understand exact science, which is about quantity and numbers, or they had no deep philosophical background to demonstrate clearly what it takes to bring science and faith on the same ground.[155] Jaki defines science consistently as a quantitative study of quantitative things in motion. He was inspired by the biblical text from the book of Wisdom: "God arranged everything according to measure, number and weight."[156] The above quotation appears several times in many of his publications.

In this chapter we are going to use the model of Ian Barbour to collocate the arguments of Jaki in, as far as the relationship between science and religion is concerned. His attempt to explain the relationship that exist between science and religion eventually gave birth to *The Impassable Divide* (2008) and *The Mirage of the Conflict Between Science and Religion* (2009). Many scholars have endeavoured, influenced by their own backgrounds and time, to demonstrate a relationship that exist between science and religion, some have taken the historical approach based on historical evidences, others have taken the epistemological approach basing on philosophical arguments taking science and religion as forms of knowledge and others tackled it from purely theological level. These attempts have yielded different results depending on the honesty and temperaments of their authors.

I have chosen to use the model of Ian Barbour to discuss Jaki's relationship between science and religion mainly because of the following reasons: First of all, because Barbour offers a general classification of how

155 For details of Jaki's treatment on numbers see S. L. Jaki, *The Drama of Quantities,* Real View Books, Port Huron (MI) 2005.

156 S. L. Jaki, *God and the Cosmologists,* Real View Books, Fraser (MI) 1998, 110. Wisdom 11:20.

in the course of history scholars have sought to explain the relationship between science and religion. Secondly, according to me, the classification of Barbour in the four models is comprehensive and coherent enough to cover the areas of Jaki's arguments that I would like to discuss. In the third place, I have chosen the models of Barbour because many other approaches fall within the four models of Barbour and some actually cut across them. Many modern models are either interpretation of Barbour or some sort of a reaction to it or a springboard to new interpretations. Lastly, a secondary reason that may not be ignored is the fact that both Jaki and Barbour had the privilege of participating at the Gifford Lectures: Jaki from 1975 to 1976, which led to the eventual publication of *The Road of Science and the Ways to God,* and Barbour from 1989 to 1991, from which he first published *Religion in an Age of Science* (1990). In my opinion, the two are in some way academically related. There are evidences to this relationship that I do not want to venture into right now but quote just one instance when Jaki asserts:

> In so far as the relationship between science and religion in an age of science can say on both counts, the promoters of religion in an age of science should think twice before taking full stock of the rights of reason if they are to unmask the wrongs done to reason in the name of ideologies grafted on to science.[157]

That Jaki explicitly refers to *Religion in the Age of Science* (1990), goes a long way to demonstrate the intellectual relationship that in some sense existed between the two. In the above quote, he is however critical of the fact that what creates confusions in the understanding of the relationship between science and religion is the mixture of science and religion with ideologies. One needs a mind free of ideologies and honest to the search of truth to establish a good understanding of science and religion and thus a proper relationship between them. Ideologies are stumbling blocks because their content are taken without scrutiny and thus devoid of the rigour reason demands in matters spiny like the rapport between science and religion.

---

157 S. L Jaki, *The Mirage of Conflict Between Science and Religion,* Real View Books, New Hope (KY) 2009, 72.

I will also discuss the cosmological arguments raised by Jaki in as far as he seeks to firmly root the relationship between science and religion in the use of good reason and philosophy. I will discuss the role played by philosophy as a bridge between science and religion, as far as a bridge can only be built thanks to the two river banks.

2.2 The Models of Ian Graeme Barbour of the Relationships between Science and Religion

Ian Graeme Barbour is one scholar who has endeavoured in several of his publications to develop a clear picture of the relationship between science and religion over the course of history beginning with his early work *Religion in the Age of Science* (1990) that eventually was developed and expanded into the latest publication *Religion and Science: Historical and Contemporary Issues* (1997).[158] In discussing the possible relationships between science and religion Barbour distinguishes the two camps using the popular opinion in the following terms:

> Science seems to provide the only reliable path to knowledge. Many people view science as objective, universal, rational, and based on solid observational evidence. Religion, by contrast, seems to be subjective, parochial, emotional, and based on traditions or authorities that disagree with each other.[159]

The above citation refers to the opinion generally held by the positivists and sets the tone of the discussion of the four models proposed by Barbour. I am going to present in a synthetic way the four models of Barbour for the purpose of my point of interest, since our concern is to demonstrate Jaki's thinking about the same and not discussing to details of Barbour's philosophy. The four models according to Barbour are: conflict, independence, dialogue and integration.

---

158 Cf. I.G. Barbour, *Religion and Science. Historical and Contemporary Issues*, Harper & Row, San Francisco (CA) 1997. Id., "Ways of Relating Science and Theology," in R. J. Russell – W. R. Stoeger – G. V. Coyne (eds.), *Physics, Philosophy and Theology. A Common Quest for Understanding*, Vatican City 1988, 21–48.

159 Id., *Religion and Science*, 77.

2.2.1 Conflict Model

The conflict model seems to be one of the most famous models of the relationship between science and religion that many scholars have propagated over the history of philosophy, science and religion. Some scholars right away see a conflict between the two based on their understanding of the cosmos and what the Bible says about the origin of things. Conflict is made inevitable when the two are put together without the recognition of their differences. Even when finding of science like the 'big bang' looks very similar or even supports biblical account, it cannot be taken as a matter of conformity to explain the biblical context of the origin of the universe. The big bang begins from existence while creation begins from nothing. Trying to merge the two will definitely brew conflicts. Some scholars describe the conflict model as the "warfare model" which, according to Barbour, comes out of the fact that they use the media as a war front. Sam Harris, Richard Dawkins, Daniel Dennett and Christopher Hitchens are typical examples of media frontliners. Barbour summarises the core of the conflict as: "Scientific materialism is at the opposite end of the theological spectrum from biblical literalism."[160] The core of this conflict is rooted in the encroachment of boundaries of theology and science and the lack of intellectual honesty that exist among the various proponents of both science and religion. Science on one hand claims to be the only reliable source of knowledge of the natural world, but scientist don't stop at that but rather advance to discuss things that fall beyond the material world and, thus, getting into direct confrontation with the theologian. Here, we are faced with epistemological reductionism according to Barbour. Positivism and the Vienna Circle go a long way to explain the font of this conflict in the history of philosophy. Jacques Monod *Chance and Necessity* (1972) and Edward Osborn Wilson *Sociobiology* (1975) are propagators of such a material reductionism.[161]

In the conflict model according to Barbour, the role played by biblical literalism, right from the medieval time that had its own method of studies

---

160 *Ibidem.*, p. 78.

161 J. Monod, *Chance and Necessity*, New York, Vintage Books, 1972, 180; E. O. Wilson, *Sociobiology. The New Synthesis,* Harvard University Press, Cambridge (MA) 1975, 4.

to the modern times, cannot be ignored.[162] How much of what the Bible says about the universe should be taken literally and how much of it allegorically? Among other requirements, this alone demands that one be competent in both scientific and biblical matters before even thinking of relating them. Galileo's two books, the scriptures and nature, should be taken into strict scrutiny in this discussion regardless of the merits and demerits of his time. The evolution theory of Darwin put in confrontation with Biblical account of the origins, that has taken various interpretations over the time, and the modern discussions on scientific cosmology on the age of the cosmos, are conflict areas put forward by Barbour. There are two extreme positions here, the radical separation of science and religion, as two opposing forces that cannot live together, and the illicit reduction of all things to be subject to scientific or religious explanations.

The conflict model and its approach has demonstrated the historical issues that has occurred between science and religion, but whether these conflicts could have been avoided or not is a matter that needs to be established. This model however falls into a deep form of reductionism and the disunion of the human rational operation that interrogates both science and religion. There is then an underlying attitude that one can either be a scientist or a person of faith but not both. If man is however the same person, who by the use of his intellect and reason interrogates nature both from a scientific and religious perspective, then a better model should be sought to explain the relationship that exists between science and religion.

2.2.2 Independence Model

The independence model is the radical view that each of the two camps are independent from each other and there is no relationship between them. It is a counter argument to the conflict model and denies that there is any conflict between science and religion because they have no relationship to cause that tension. Barbour says, "This separation into watertight compartments is motivated not simply by the desire to avoid unnecessary conflicts, but also by the desire to be faithful to the distinctive character of each area of life

162 Cf. I. G. Barbour, *Religion and Science*, 82–84.

and thought."[163] Religion and Science are independent because they both have different methods and different formal objects, as science deals with the material composition of natural things and how they function, religion deals with their spiritual aspects and their purposes. The independent model took different tones in the course of history like contrasting spirit and matter, revealed truth and scientific discovery. Renowned scholars who held this view include Karl Barth, with a clear contrast between the divine and human. The German theologian Rudolf Bultmann, with religious existentialism, asserts that "theological formulations must be statements about the transformation of human life by a new understanding of personal existence. Such affirmations have no connection with scientific theories about external events in the impersonal order of a law-abiding world."[164] Other scholars whose works were considered by Barbour are Langdon Gilkey and Thomas Torrance.

According to Barbour, "an even more effective way of separating science and religion is to interpret them as languages that are unrelated because their functions are totally different."[165] The language of religion cannot be conformed with the language of science, since they all address different aspects of the human existence and the cosmos. According to Barbour, scientific language deals mainly with prediction and control, while religious language deals with the way of life. Renowned scholars who discuss language include George Lindbeck who, in *The Nature of Doctrine* (1984), argues that religious language cannot be used by science in the same way. Barbour acknowledges that the independent model is a good step towards the right direction but it is not sufficient. He concludes that "we cannot remain content with a plurality of unrelated languages if they are languages about the same world. If we seek a coherent interpretation of all experience, we cannot avoid the search for a unified world view."[166] According to Barbour, it is impossible in real life to have a complete independence in any human practice and therefore there must be a better way for a fruitful

163 *Ibidem.*, p. 84.
164 Cf. R. Bultmann, *Jesus Christ and Mythology*, Charles Scribners Sons, New York, 1958.
165 I. G. Barbour, *Religion and Science*, 87.
166 *Ibidem.*, p. 89.

relation, because in real life wholeness and interconnectedness come first; from which we later develop particular areas of expertise. This model also falls short of the acknowledgment of the capability of the human mind to unite in a coherent body of knowledge of separate things. There should therefore be a possibility of dialogue between science and religion.

2.2.3 Dialogue Model

This is a moderate view that seeks to bring into dialogue science and religion. Defenders of this view assert that science and religion should be in dialogue not only for the good of the human person, but also because they all have nature as their primitive source and point of departure. According to Barbour, dialogue model agrees that there are direct interactions between religion and science taking into consideration boundary and methods. He argues that as far as the boundary of discipline is concerned, science has developed from ancient cultures and religions. This argument is based on the fact that the universe is orderly and that its existence must have some explanations that science looks for in nature itself and religion, beyond nature. The shortfall of ancient cultures was that many of them took nature to be divine. Barbour quotes a strongly held position of Alfred North Whitehead in *Science and the Modern World* (1925), that "while nature is real and good, it is not itself divine, as many ancient cultures held. Humans are therefore permitted to experiment on nature."[167] The freedom to experiment on nature however, is held by Jaki in *The Road to Science and the Ways to God* (1978), as coming directly from the Christian conviction that the universe was created by God who is outside creation itself. Barbour however quickly reacts to Jaki's position without a mention of him, that the Christians' role in the development of science should not be overstated. He argues that "when modern science did develop in Europe, it was aided by humanistic interest of the renaissance, the growth of crafts, trade, and commerce and new patterns of leisure and education. Yet it does appear that the idea of creation gave a religious legitimacy to scientific inquiry."[168]

167 *Ibidem.*, p. 90.
168 *Ibidem.*

Barbour is careful and uses the expression that "it appears" to avoid, most probably, a previously stated caution that the role of Christianity should not be overstated. Jaki would out rightly argue in many of his works that a viable birth of science was only possible in the Christian West, as we have seen in the previous chapter. Being a scientist, he argues that the science of quantities does not admit of a "more or less" situation and that is why he calls it exact science. We shall discuss his position on the relationship between science and religion.

Barbour ultimately wants to demonstrate that there has been and there will be moments of dialogue that each side will outgrow and move back to their camps after having left impacts on each other. "I believe the case for the historical contribution of Christianity to the rise of science is convincing. But once science was well established, its own success was sufficient justification for many scientists, without the need for religious legitimation."[169] For Barbour dialogue is possible when no side over emphasizes its importance. He acknowledges that science raises religious questions in the course of their investigations that it cannot provide answer for. He goes ahead and discusses the views of Thomas Torrance in *Theological Science* (1969), Wolfhart Pannenberg in *Theology and the Philosophy of science* (1976), and catholic authors like Ernan McMullin (1924–2011), the German Jesuit Karl Rahner (1904–1984) and the American theologian David Tracy (1939).

As far as methodological parallels are concerned, Barbour argues that the parallels between science and religion, as far as their methods of investigations are concerned, are not that of separation but degrees: "The scientific criteria of coherence, comprehensiveness and fruitfulness have their parallels in religious thoughts."[170] He also discusses the views of Thomas Kuhn's *The Structure of Scientific Revolution* (1970), in which Kuhn argues that the scientific community share scientific paradigms while the religious community share the religious paradigms in different degrees. Barbour also discusses the views of Michael Polany who tackles issues related to faith and reason and sees dialogue between science and religion

169 *Ibidem.*
170 *Ibidem.*, p. 93.

made possible by a personal participation of the knower in all knowledge. Modern authors discussed by Barbour include John Polkinhorne, *One world: The interaction of Science and Theology* (1987) and Holmes Rolston, *Science and Religion, A critical Survey* (1987).[171] I will not discuss the details of their views presented by Barbour because they fall out of the scope of this work. Can the possibility of dialogue also allow the possibility of an integration?

2.2.4 Integration Model

According to Barbour some authors agree that some form of integration is possible because as science investigates into nature, it can help throw light to religious truths that can eventually be used in reformulating doctrine. This however, does not mean that science provides grounds for doctrines but merely that science can give details that can help religion to articulate well matters of doctrine. His arguments encompass three areas: natural theology, theology of nature and systematic synthesis. On natural theology Barbour argues that the arguments for the existence of God, for which Thomas Aquinas is well known, are derived from human reason other than historical revelation or religious experience. These forms of arguments are sometimes directly drawn from scientific evidences. Reason is at the core of every argument arising from nature and that same reason is also necessary for scientific investigations. This alludes already to some common point between the two. On the theology of nature, Barbour argues that there is some overlap between the concerns of science and religion. For instance, "the doctrine of creation, providence and human nature are affected by the finding of science."[172] This means that there is some intersection between science and religion that begs for another method of investigation. He cites the arguments of Arthur Peacock in *Imitation of Reality* (1984), that God creates "in and through" the process of the natural world that science unveils, as typical of the integration approach. For Barbour, to arrive at a systematic synthesis, "we need philosophical categories to help us unify scientific and theological assertions in a more systematic way."[173] He

171 *Ibidem.,* pp. 94–98.
172 *Ibidem.,* p.100.
173 *Ibidem.,* p. 101.

asserts that authors like Teilhard De Chardin in *The Phenomena of Man* (1959), have tried this integration not without creating a lot of theological and scientific confusions. Barbour suggest that the views of Chardin be "viewed as a synthesis of scientific ideas with religious ideas derived from Christian tradition and experience."[174] Chardin's biggest mistake was the temptation to use scientific facts and findings to justify religious beliefs, in a way of amalgamating them, falling into absurd conformism. He presents evolution as a necessity that cannot be avoided to develop any theory or system. Teilhard limits every event and understanding to the framework of evolution and therefore, enters into the difficulty of explaining instant existence that disobeys any linear progress.

According to Barbour, "a more systematic integration can occur if both science and religion contribute to a coherent world view elaborated in a comprehensive metaphysics."[175] This provokes a quick question why there is need for such an integration but also gives a good leap into that comprehensive metaphysics that he proposes, that demands then an intervention of another field of studies. To him, a philosopher should take the lead even when theologians and scientists can benefit from the integration, so that it serves as a melting point for common reflection. Barbour suggest that there is need to articulate the theology of nature, which deals with the overlapping elements of science and religion, in order to achieve a good integration. This is because, according to him, from it "we will have to draw from religious and scientific sources." Barbour argues that in pursuing integration using comprehensive metaphysics:

> There are dangers if either scientific or religious ideas are distorted to fit preconceived synthesis that claims to encompass all reality. We must always keep in mind the rich diversity of our experience. We distort it if we cut it up into separate realms or watertight compartments, but we also distort it if we force it into a neat intellectual system. A coherent vision of reality can still allow for the distinctiveness of differing types of experience.[176]

Barbour's stand here will usher us into the collocation of Jaki's thoughts in the four models. Jaki argues that science and religion are forever

174 *Ibidem.*
175 *Ibidem.*, p. 103.
176 *Ibidem.*, p. 105.

conceptually distinct and there should be no attempt to fuse them. The mind can hover over in search of understanding but the two domains remain separate. The analysis of Gilson would fit very well to explain why distinct elements cannot be put together without creating confusion, he thus writes, "only distinct things can be united; if you attempt to blend them, you inevitably lose them in what is not union, but confusion."[177] My interest is to underscore the power of the mind to unify various domains of the human intellectual operation into a coherent body of knowledge without having to fuse them into one. A superficial consideration of Jaki's arguments may make it appear as if he is an advocate of the independence model but he makes the intellectual integration crystal clear when he asserts that both sides are real sides and deal with the same reality in two different ways and it does not matter on which side one finds himself.[178] We shall discuss this in depth in the next section.

2.3 Collocating Jaki's Thoughts in the Models of Barbour

Jaki does not explicitly refer in his discussion of the relationship between science and religion to the four models of Barbour. He does not also classify his arguments for the relationship between science and religion into some model. It is my attempt in this exposure, dictated by one of my objectives, which is, the philosophical reconstruction of Jaki's thoughts, to collocate his arguments for the relationship between science and religion in the existing classical models of Barbour. If this will be made clear at the end of this chapter, it will be a great step towards a philosophical reconstruction of Jaki's historical, theological and philosophical thoughts. The major preoccupation of Jaki was the reconstruction of the history of science in order to demonstrate the critical moments of its development into the science that we now have. Jaki himself confirms the necessity of our objective, when he says, referring to his research on Duhem:

177 E. Gilson, *The Unity of Philosophical Experience,* Charles Scribner's Sons, New York 1937, 50.

178 Cf. S. L. Jaki, *The Impassable Divide,* Real View Books, New Hope (KY) 2008, 99.

> For the time being, my mind was still more that of a historian than that of a philosopher. Whatever philosophical point I tried to make I still did so by expressing it as contained in this or that phase of the history of science. [...] The historical approach still played the principal role in what may be considered my first major work in philosophy and not merely in the philosophy of science.[179]

I should however say that Jaki did not wean himself of the historical approach throughout his carrier even when he was called to run courses in philosophy. This is justifiable because the reconstruction of the history of science requires philosophy if it is to take intellectual moments that science made the greatest stride. In as far as the classification of the relationship between science and religion is concerned, examining our author's arguments, we note a clear awareness of the existence of the various classifications of the said relationship. In establishing his vision, he systematically eliminates the various existing models and takes his personal stand as a scientist, theologian and philosopher without making a preference for any model, the task that I am set to accomplish.

2.3.1 Collocating Jaki's Thoughts in Conflict and Independence Models

Jaki explicitly discusses the inexistence of a conflict between science and religion and thus eliminates the possibility of collocating his thoughts in the conflict model but also in the same wake in the independence model of Barbour. In *The Mirage of Conflict between Science and Religion* (2009), Jaki argues that it can only be clear that there is no conflict between science and religion when the two are strictly defined. By defining well science and religion, we set the limits and boundaries of their competence but also purge them of what they are not. Exact science deals strictly with the measurement of quantities of things in motion expressed in integers while religion deals with spiritual entities that have no scientific measurement. According to Jaki, science does not admit of a predication of more or less since it deals with numbers, as long stated by Aristotle in the Categories "where the Stagirite notes the impossibility of predicating 'more or less'

179 Id., *Mind's Matter: An Intellectual Autobiography*, William Eerdmans Publishing Company, Grand Rapids (MI) 2002, 81.

of integers, whereas the term is freely and rightly used with regards to all words belonging to all categories."[180] Jaki continues to argue that the Catholic religion in the propositions and dogma of faith has nothing to strictly do with integers even when the doctrine of the Holy Trinity is expressed as one God in three persons.[181] Stephen Barr holds the same view and states emphatically, "It is therefore perfectly true to say that science and religion are not in collision."[182] The collision that should escalate into conflict is therefore eliminated by the fact that the two deal with different realities that are irreducible to each other.

It is apparent that religious faith cannot be expressed in numbers, because it is immeasurable and yet reasonable. For him, there is a fine line between religion and science. Religion follows a two-way path that begins from God's love for mankind and man's response to God's love in faith, while science on the other hand follows a one-way path that becomes progressively narrower and ends up in formulae expressed in numbers that are irrelevant to religion.[183] The more science progresses in its specialisation, the more it loses the sense of the whole and becomes strictly specialised in parts. This partial treatment of the totality of reality defines at the same time its limits and the illegitimacy of any universal claim by the scientists. Strict science is not relational but factual and functional, and consequently, cannot be in conflict with religion, that is based on ontological truths revealed or solemnly proclaimed in the form of a dogma. Jaki acknowledges the fact that there have been clashes between the catholic religion and science but not necessarily a conflict.[184] He argues that the case of Galileo was one instance of a clash, because "Galileo failed to provide the theory of motion which

180 Id., *The Mirage of Conflict Between Science and Religion*, viii, 13, 45.

181 Cf. *Ibidem.*

182 S. M. Barr, *Modern Physics and Ancient Faith*, 1.

183 Cf., S. L. Jaki, *Lectures in the Vatican Gardens*, Real View Books, New Hope (KY) 2009, 185.

184 Jaki argues that Catholicism is the only religion worthy of the name. He argues from the etymology of the term 'religion' from the Latin *re-legare* which means retying or reconnecting a broken relationship. For him only the catholic religion with the dogma of the original sin seeks to reconnect a broken relationship.

the heliocentric system demanded."[185] It was because Galileo could not prove quantitatively what he was claiming.

> Thus, without a clear quantitative challenge, there arose a conflict between science and religion, the first major conflict between the two. The challenge was clearly incomplete on the part of science, but it would have been non-human to expect believing scientists to wait until all was clear. In actual life, matters often come to clash before there is full justification for it. Some historians of science even say that science should thank the church for having forced Galileo to return to his strictly scientific pursuit from being a propagandist for Copernicanism.[186]

We note from the above quotation how much the term 'conflict' has been entrenched as a terminology in the description of the relationship between science and religion, that Jaki in denying the inexistence of conflict uses the same term conflict. He however argues that this apparent conflict was because Galileo did not have at that time the quantitative evidences to challenge the existing belief. He asserts: "Still, Galileo could not provide evidence on behalf of his crucial contention, namely, that the earth rotated on its axis and moved around the sun."[187] The point of conflict is that Galileo abandoned the scientific method and somehow believed without an experimental prove that the earth rotated around the sun. Even if this turned out later to be proven, such a proof was not available at the time of the controversy. According to Jaki, the Catholic Church on the other end insisted that science should be science, with all its methods and limits, and should have proofs for its claims. The clash, therefore, came because the Catholic Church did not give time for Galileo's science to mature, basing on the already divergent opinions on Copernicanism, and Galileo did not have the patience to wait until he had quantitative evidences to support his claim. If there were a conflict, then it was not between science and religion but between the Church leaders with their religious understanding on one hand and Galileo with his scientific propositions on the other hand. The fact that science continued to flourish both in Italy and the rest of Europe is a clear testimony that the apparent conflict was between Galileo and the ecclesiastical authorities who insisted that science be science with

---

185 Id., *The Mirage of Conflict between Science and Religion*, 1.
186 Id., *Questions on Science and Religion*, 101–102.
187 *Ibidem.*, p. 104.

quantitative proofs. Galileo was involved both in science and ideological advocacy at one point of the contention. The clash was therefore between what nature through speculation portrays and what science through quantitative measurement could establish.

Jaki argues that the case of Galileo could have been avoided if the Church authorities had paid attention to the general rules of the interpretation of the Bible and science proposed by St. Augustine that reason should take precedence where the Bible seems to say something with a different connotation. Augustine long suggested when a biblical text can be taken literally and when it should be subjected to other forms of interpretations basing on evidences and the context in which the church finds itself. It is apparently clear that Galileo was also aware of this provision by St. Augustine. Jaki quotes the famous saying attributed to Galileo interpreting St. Augustine that religion is about going to heaven and not about how heaven goes.[188] How heaven goes is the business of astrophysics and scientist interested in its motion and how to go to heaven is the business of religion that has nothing to do with quantities and their measurements.

In communicating their contents, religion and science have completely different targets, this is due to the fact that they have different material and formal objects even when these objects are real existence. The material object refers to the "material" studied, that is, that which is put under investigation as the object of its research.[189] It is the point of departure of a specific discipline and in our case that of science and religion. While the formal object of a discipline is the point of view from which the material object is studied and the methods used to carry out an investigation. Every

---

188 Id., *The Impassable Divide,* 71. St. Augustine's *De Genesi ad litteram* about the Holy Spirit's intent in inspiring the Bible – which culminates in the statement: *Spiritus Dei noluisse ista docere homines nulli saluti profutura.* ("The Spirit of God did not want to teach people things that would be of no help to their salvation."). Cf. G. Galilei, *Lettere,* Einaudi, Torino, 1978, 128–135.

189 The term 'material' should not be taken in a materialistic sense. We can for example assert that the soul is the material object of the philosophy of psychology, even when the soul is not material in form, or when we talk of God as the material object of the various theological disciplines, we do not want to think or say that God is material but that in studying theology, the object of studies is God in relation to His creation.

science therefore has its material and formal object that distinguishes them. In fact, it is the method used and the consequential formal perspective that defines the epistemological value of every science. In demonstrating that there was a clash and not a conflict between science and religion, Jaki brings in the fact that science and religion have two different formal objects and use different methods to acquire knowledge about their material objects. He argues that, while religion is concerned with the same concrete things provided by nature with completely different point of view, its formal object are spiritual concepts relating to human existence. While science on the other hand has as its formal object the composition of matter and how it functions. They ask different questions on the same reality and therefore acquire knowledge relative to their fields. The concept of the universe as an abstraction of an aggregate of material realities cannot form a formal object of science. To substantiate, Jaki asserts: "It may surprise many even within scientific observations realms to hear that treating material totality cannot be the object of scientific observation. Science as such can never prove that there is a universe, a strict overreaching totality."[190] Science as strictly science deals with very small aspects of even individual realities and therefore it is plausible that it has no tool whatsoever to delve into the discourse about the universe or the human soul which is the basis of man's free will and sense of purpose. As philosophically we can talk about the universe as an aggregate of all things that exists and operates together, scientifically we can only talk of sets of things and how they act similarly or in different modes.

Following different paths of investigations, Jaki argues that "a most precious truth is that there can be no real conflict between two fields, one of which, science, is pivoted in numbers, whereas the other, religion, has no essential use for them."[191] Jaki concludes with these words: "There may be clashes between science and religion, or rather between some religionists and some scientists, but no irresolvable fundamental conflict. [...] But the task of both must be carried out if man is to remain sane in a welter of change which operates with an ever more powerful dynamism."[192] The

190 *Ibidem.*, p. 3.
191 *Ibidem.*, p. 9.
192 *Ibidem.*, p. 10.

above quotation does not only eliminate the possibility of conflict between science and religion but also opens up to a possibility of another type of relationship that should exists between the two. The apparent conflicts are resolvable if both sides accept to remain within their competences and employ a sincere use of reason. Man has no option but to engage in both science and religion. It is also clear that Jaki wants to point towards those doing religion without a strong philosophical foundation as the cause of the apparent conflict. There are those he refers to as religionists as to distinguish their conflict with science as technically not with religion itself. So, there can be real conflict between religionists and scientist because of the difference in their capacities to comprehend the relationship that should be between science and religion.

If man is to be that rational animal with the power of self-recognition, and if this same man using his intellect must understand his environment and think beyond it, then there should be that internal harmony in him that sees separate things as separate and related things as related. The scientist and the faithful can exist in the same man without contradiction. It is another thing whether the individual would perceive that the co-existence in question is due to the conceptual irreducibility of the two perspectives.[193] That self-awareness and restrained is cultivated by seeking to know in depth one's own area of specialization but never in isolation.

In discussing the relationship between science and religion, the problem of method will always return, because it is the one that decides the path to be followed, and the reasons for apparent conflict is stipulated by Jaki thus:

> Scientist no less than theologians can generate false conflicts between their respective fields by ignoring the limits of their methods. Scientist can be as much overimpressed by the power of their method as theologians can be tempted to attribute more to the eternal truths about salvation than these were meant to deliver.[194]

Science has over time efficiently responded to human practical needs and has succeeded in putting in very sophisticated technologies. It seems that it answers most human quest, quickly, efficiently and sometimes permanently. This in itself is positive but the risk is that scientists begin to think that

---

193 Cf. *Ibidem.*, p. 13.

194 *Ibidem.*, p. 18.

science can do it all and venture into answering questions that do not regard them. It can then be argued that the employment of rigorous reason and the intellectual power should be the demarcating tools for the boundary between the two. This is also true for the theologian who because of the confidence in his faith risk to obscure the glaring achievements of science. I would assert that it is not enough to ask metaphysical questions but also equally important to engage in rational discourse about the question raised and the best candidate in that field is the philosopher. Faith and science can exist in the same person without contradiction as long as the individual reasonably understands them. This implies that ontological questions can be asked by both the scientist and the theologian without any of them ever exhausting the truth about it but also without necessarily having to give an answer to them. To avoid conflicts, it would be prudent not to attempt to give authoritative answers in areas that one has no competence, but merely opinions, for which everyone has a right. Such opinions should however be taken as so, bearing only the value it should have within its isolation.

Jaki continues to argue that "man's registering of reality, his acting freely and for a purpose, his sense of morality – all these fundamental facets remain forever outside the methods of science and eliminates the possibility of a fundamental conflict between science and religion."[195] The freedom of man that determines his choices, which are judged as good or bad, just or unjust, and his acting for a particular reason, cannot be subjected to quantitative measurement and therefore have no grounds for conflicts with a quantitatively measurable things. That may explain why "the road or method of physics is open in every direction, right, left, up and down, but blocked where the object of reflection offers nothing to measure."[196] These all lie outside the science of numbers. However, according to Jaki, "numbers cannot give more than quantitative relations, but when tied to physical reality they reveal about the latter a property of it which shows that it owes its existence to a creative, selective act by God the Father."[197] We note here that examining physical reality the scientist is rationally drown to ask questions about what caused it to be and cannot attribute one cause to

195 *Ibidem.*, p. 19.
196 *Ibidem.*, p. 20.
197 *Ibidem.*, p. 51.

another *ad infinitum*. There must be the first cause which is responsible for all other secondary causes and according to St. Thomas, it is what everyone calls God. Science by its methods cannot arrive to this conclusion without the mediation of the philosopher and yet it continues to point to something that transcends it.

Jaki tackles directly the issue of scientific materialism and biblical literalism showing how it has defined the relationship between science and religion in the course of history. According to him, there are infinite evidences in the Bible of the creation of the universe. It is however a grave mistake to read the book of Genesis as a science textbook. This seems to be the greatest mistake made by many historians of science and scientists. Scientists who try to disprove the account of creation in Genesis have either not understood what they should be doing or have nothing to do at all. He asserts that the world was created out of nothing as long attested by the Christian dogma.[198] It would seem apparent that the fact of the creation of the universe out of nothing is proved by the inability of the scientists to arrive at that moment of creation. This is because arriving at it is actually arriving at nothing. We can actually conclude that science has proved the creation out of nothing by a negative movement asserting the impossibility of evidence. He further argues that there can be no scientific proof of creation out of nothing for a simple reason that nothingness cannot be observed or measured. This implies that beyond physical reality lies a nonphysical one that can only be arrived at by inference from the concrete physical existence.

According to Jaki, science cannot competently have for its object the creation of the universe out of nothing or the determination of its eternity or of its multiplicity. The biblical language cannot be used with the same meaning in scientific research and experimentations. The Bible has its own way of interpretation of even what appears as facts. That the Bible has a different end and should be accepted wholly as it is. A selective acceptance militates against science by its very nature and against reason by its very

---

198 Cf. Id., *Bible and Science*, Christendom Press, Front Royal, (VA) 1996. Cf. S. Barr, *The Believing Scientist*, William B. Eerdmans Publishing Company, Grand Rapids (MI) 2016, 181. He defines creation as "that timeless act whereby God holds all things in existence. It is not an alternative to natural theories of origin or natural explanations of change."

operation. The universe is an object of philosophy and ultimately metaphysics. According to Jaki, metaphysical reasoning (philosophy) alone can assure the scientific cosmologists that there is a coherent totality of material things. It is evident therefore that science held within the limits that appreciates philosophy rationally leads to theology and hopefully faith. It is a rational conclusion from evidences that do not contradict reason. It is the conviction in the created nature that we derive its rationality which consequently leads to the trust in the rationality of a created mind.

It is probably worthwhile to conclude this section with the assertion of Jaki that demonstrate the interplay of rationality as derived from creation of the universe and how it gives the possibility of doing a rational physical science.

> Modern cosmology too will prove itself immune to the disease of solipsism and anthropocentrism which trapped some of the best cosmologists of our time. But without a reference to the doctrine of creation, which implies both the rationality and contingency of the universe and the matching rationality and contingency of *homo faber,* there will be no consistent explanation of the scientific method, a most delicate interplay of experimentation and reasoning, nor will there be a consistent correlation of cosmological theories, however powerful, with the cosmos, the coherence of all real contingent things.[199]

The fact that creation presents itself to human mind and affects it is an evidence of its rationality but more importantly is that it provokes in the human person the desire to know it, because he is capable of reason, the desire to investigate into it and develop theories to understand it. This goes back to the Aristotelian assertion that all men by nature seek to know and this knowledge in its universal form is philosophy. Science therefore does not require experimentation alone but an interplay of a rational discourse and experiments. Experiments reveal facts about the cosmos that is organised by human reason into theories that explain the underlying principles behind a particular cosmic phenomenon. These theories and principles are picked up by philosophy to produce an epistemology of the universe that serves to make religion reasonable to the human mind. Being reasonable, the scientist has to bear in mind that every scientific investigation has consequences when not handled in the most rational way possible. These

---

199 Id., *Cosmos and Creator*, Regnery Gateway, Chicago (IL) 1980, 139.

consequences are either good and therefore ethical or bad and unethical. For the scientist to define his actions as good or bad, he requires the mediation of the philosopher who does not only look at goodness in a religious term but in a rational and coherent way. Belief that the universe is created is an incentive to the scientist to investigate into it in order to decrypt its content not only as an adventure but more so as a participation into the creative work of its author. Faith helps the scientists to collocate creation outside the creator and therefore subject to his investigation. He however needs the philosopher for him to transcend material things into metaphysical realities. Jaki, discussing the importance of faith in science asserts:

> Faith in the intelligibility of nature, faith in the possibility of reaching the outer world with our concepts, faith in the existence of a universe structured according to definite patterns and revealing orderly arrangements in its entirety and its details. As Einstein expressed it: 'Belief in an external world independent of the perceiving subject, is the basis of all natural science.' To him the nature of such faith was genuinely metaphysical that he did not refrain from putting it in the sphere of religious beliefs.[200]

It can be argued that the faith in the stability and reasonableness of nature by the scientist may not be the same with that faith that one has in believing in God. It remains a fact that the scientist believes in something called order and stability that allows his operation and that lies outside his ability to measure, that makes measurement possible. It is however possible with the mediation of philosophy, using metaphysics that one's conviction in the reality of nature makes the religious belief in the same, plausible and reasonable. This calls for the recognition of faith as a foundation for scientific work.[201]

Anthropocentrism is a stumbling block here because it seeks the cause of things strictly in man's ability to decipher them strictly in a materialistic way and yet so much lies outside the material world and man's abilities. A sound discourse about the universe will therefore take the form of a metaphysical realism that does not build a castle in the air but that has the cosmos as an indispensable stepping stone for a rational recognition of a creator. Jaki asserts that for this to happen, one must first say 'creator' before saying

200 Id., *The Relevance of Physics,* 345.
201 *Ibidem.*

cosmos.[202] Using reason one is enabled to do cosmological science and ask metaphysical questions which by its very nature already falls outside the scope and competence of cosmological science, there is therefore need for philosophy to act as a bridge between cosmological reality and metaphysical questions provoked by it. It is the role of philosophy to put together the underlying universal principles behind physical realities that links to metaphysical questions in religion.

For Jaki therefore, there is no conflict between science and religion because: in the first instance, their strict definitions do not warrant conflict because it separates them; secondly, the two have different material objects and therefore different methods of investigations within the boundary that separates them and that should be respected; thirdly, the one who seeks the knowledge of science and knowledge of God is the same person who possesses both ontological and quantitative capacities, it is therefore unthinkable that reason can be in conflict with itself in the same person without alienating him. Jaki asserts: "The air would be much clearer about a number of topics, including the subject of the relation of science and religion, if all parties in a dispute would specify their antecedent principles."[203] Although from what has been said so far Jaki seems more to suggest than to affirm this argument, as we shall see later in this chapter that he also proposes arguments to support the mediating role between science and religion that can be achieved by good philosophy. His arguments are based on the fact that both science and religion carry out investigations on concrete reality that alludes to the unity of human rationality. After having seen that we cannot collocate Jaki's thoughts in the conflict model, we would like to examine his thinking in relation to the Independence model.

In arguing for the non-existence of conflict between science and religion, Jaki also eliminates the possibility of cataloguing his thoughts in the independence model even when it is apparent at first sight that his arguments draw him close to it. Whereas science and religion use different methods and have different formal objects, they have two things in common that makes a watertight independence impossible: both science and religion use

---

202 Id., *Cosmos and Creator,* 141.
203 Id., *The Mirage of Conflict Between Science and Religion,* 64.

reason to arrive to their conclusions departing from their primary source, which is concrete natural things, notwithstanding revelation as an added advantage for the Christian faith. I must assert that even revelation does not go against reason as some holders of scientism would wish to portray. Since the human reason is only one and cannot be divided into religious and scientific reason, there is a relationship between science and religion that makes the independence model not only a contradiction in terms but also an attempt to alienate man from his rational faculty. Jaki demonstrate this common rationality as directly coming from the creator whose act of creation makes the whole of nature rational and therefore any scientific discourse about it must be rationally related. The rationality of creation is derived from basically two fundamental realities, purpose and order in creation. Nature seems at all times to follow some law that makes it orderly and directed towards an end. This then implies an intelligence behind it that is called by religion creator. Since there is this order and purpose, it follows that the universe is a product of an intelligence that is superior to it. It is this reason manifest in creation that makes it a rational product as opposed to randomness. Jaki writes:

> Human reason can use consideration about things visible to make the inference that there has to be primordial reason or rationality, which is so primordial that in it, truth and existence co-exist, indeed are the same. Therefore, such a principle is also the ultimate source of all existence. But since that source, or God in short, is reason absolute, He can create only such entities as are marked with rationality at least in the form of having consistency in their properties.[204]

The quotation above about the creator goes a long way in demonstrating the certainty and rationality of natural existence. I would like to note that natural theology and science is possible because there is a consistency in nature. If there were no consistency in nature, experimentation, which is a chief requirement of science would not be possible and neither would a rational ascent to God through nature. This same consistency eliminates chaos and chance and leads to the irrefutable ontological question why there should be something in the mode in which it is and not nothing. Faith is therefore, reasonable as opposed to a blind leap over gaps of ignorance. Jaki however warns that reason should be purged of the lack of

---

204 *Ibidem.*, p. 65.

commitment to the truth for a better understanding of a relationship that should be between science and religion. The ontological truth dealt with here alludes to the theological understanding of truth whose stable source is God – the immutable truth. Creation therefore, is true and rational by participation in the nature of its immutable creator.

From the arguments demonstrating some common factors between religion and science, it can be argued that it is true that the two should not be in any conflict if they are clearly defined, but also that the two cannot thrive in the independence model. If science and religion are not contradictory, then they cannot conflict but we want to note that the absence of conflict does not prevent Jaki's thoughts from falling within the independence model. As far as dealing with quantitative features of things in motion is concerned, science and religion are independent of each other in their specific methods but in as far as rationality is concerned, they are not. I would like to assert that the consistency in the rational properties of natural entities provides two common denominators between science and religion namely, progression by reason itself in scientific investigation and the departure from nature. With these two common denominators, science and religion, strictly speaking, cannot reasonably be independent of each other and therefore, eliminating the possibility of collocating Jaki's thinking within the independence model of Barbour. This also confirms the consistency of Jaki's thoughts with his argument that science was born as a result of Christian understanding of creation. There should therefore be a better description of the relationship between science and religion that we shall discuss in the next section.

2.3.3 Collocating Jaki's Thoughts in Dialogue and Integration Models

By the very insistence that science was born within the later medieval Christian times, Jaki already establishes a strong relationship between science and religion, if he is to be consistent in his thoughts. Deductively, this relationship can neither be of conflict nor independence but of dialogue and harmonic integration. This integration should not be taken superficially because it involves a permanent conceptual distinction of scientific and religious domains. It can be argued that Jaki conceptually distinguishes science and religion to unite them by good philosophy and natural existence, which is their common springboard. He uses the opposite of a biblical text used in the sacrament of marriage thus: "What God has separated, no one should put together." Jaki continues to argue that "Unless one considers science and religion as two distinct, separate, and mutually irreducible enterprises, one may be lured, [...] into trying to fuse the two together. The result has invariably been not a fusion but a confusion."[205] The statement 'what God has separated no man should put together' should not be taken literally but allegorically to show the impossibility of putting the two domains into one in the sense of fusing them. He constantly insisted that, there was need in history to show clearly the separation between "ontological questions about nature and motion and the indispensability of handling motion as measurement expressed in numbers that constitute exact science."[206] We have already noted that Jaki refers to science as exact science which is bluntly a reference to physics. This helps him to distinguish well science and religion.

According to Jaki, science and religion are two clearly distinct things and therefore cannot be mixed up without creating confusion. For Jaki any form of concordism will tantamount to preventing scholars from seeing the reality that the two cannot and must not be put together. He argues: "Concordism provided a blindfold against seeing that what God separated, that is, put in two mutually irreducible domains in man's mind, man should not try to fuse together."[207] This does not also imply discord between science and religion but distinctions that make them different fields. We pay attention to Jaki's details of, "irreducible domains in man's mind" which I would add in man's rational capacity, which is only one and capable of handling both facts separately. This mutual-simultaneous-distinct-existence in the human mind and reason is what I would like to call the epistemological integration of science and religion. Aristotle in the categories separated quantity which deals with unity, plurality and totality, and quality, which deals with reality, negation and limitation, as distinguishing marks of beings. Jaki takes the same argument that quantity cannot be reduced in all its parts to quality

---

205 Id., *A Late Awakening and Other Essays*, Real View Books, Port Huron (MI) 2006, 63.

206 Id., *The Mirage of Conflict Between Science and Religion*, 36.

207 Id., *The Impassable Divide*, 59.

and vice versa and since science deals with quantity of things in motion and has no mechanism to deal with quality which is a characteristic of religion, the two are conceptually and permanently distinct and should at all times be seen to be so. Jaki qualifies the distinction though that it should be in the quality of the mind and therefore it should be learnt to avoid mixing up the two. Since the distinction demands some quality of the mind, it is also clear that harmonic relationship can only be achieved by cultivating this mindset in the discipline that promotes its understanding. It is important because concrete realities have both qualitative and quantitative properties that need to be separated by knowing them.

Jaki argues that to understand that the two fields are to remain separate, one should be well versed with both and above all add on the knowledge of philosophy as true love for wisdom. He claims that even the best science cannot at any point prove that it has the ability to permeate into matters regarding universal principles which is a subject matter of philosophy. Thus he writes: "Apart from the inability of the scientific method to deal with purpose, whether individual or cosmic, true cosmic totality even today escapes measurements, which is perennial touchstone of scientific truth."[208] One can see here that purpose is at the core of religion and if it eludes science, then science has no deal with it and ultimately making statements that are purely religious in defence of scientific findings or against religion would be considered going outside the competence of science and bringing to question a sincere use of reason. Jaki even puts it more concisely, "Since the sense of purpose cannot be measured, because it cannot be expressed in grams, litres and frequency numbers, religion can never become a scientific proposition."[209] The purpose or end of the two are completely different and yet they both serve to equip the thinking being with knowledge of the physical world and what lies beyond it. Religion parts from concrete reality and revelation to God who creates everything out of nothing, which is at the same time a grantee for rationality of the universe. Science also parts rationally from a given reality but with a different interest on how it functions and not why it should be. He argues that for science nothingness has no

208 *Ibidem.,* p. 68.
209 Id., *Lectures in the Vatican Gardens,* 191.

meaning because it cannot measure nothing and yet creation out of nothing for religion is reasonable and a revealed truth.[210] Religion and science are therefore, two distinct conceptual blocks that are irreducible to each other.

To those who hold complementarity as a mode of relationship, Jaki argues that the term 'complementarity' was taken out of context from physics and applied to demonstrate a relationship that cannot be possible between science and religion.[211] According to Jaki, Niels Bohr, who is the author of the complementarity theory became the first abuser of his own theory by extending it to fields that do not concern it. The principle that "a body showed not only characteristics of a corpuscle but those of a wave," when applied to fields like religion is like saying, according to Jaki, that atheism complements theism.[212] Jaki's argument goes back to the use of language and principles that are sometimes taken directly from the scientific realm and applied directly to religion. Bohr, according to Jaki, did not pay attention that numbers could not be subjected to complementarity and stating that science complements religion would be trying to mix two separate realities that do not admit of it. Barbour has a moderate position and thinks that complementarity can be used in other disciplines by extension: he defines it as the relationship between contrasting models and concepts such as waves and particles used to describe the same thing in quantum theory.[213] This extension should however be carefully considered bearing in mind the various conditions for its application. But according to Jaki, even in physics which is science per excellence, the principle of complementarity could not be demonstrated.

> Tellingly it was not demonstrated even in physics that a piece of matter was at the same time both something really physical as well as mere wave packet. Nor was it demonstrated within parameters of relativity physics that everything was relative in the physical world. Being contradictory in itself, such a claim was indemonstrable whether by logic or by experiment.[214]

---

210 Cf. Id., *The Impassable Divide*, 70.
211 Cf. I. G. Barbour, *Religion and Science*, 167–170.
212 S. L. Jaki, *The Impassable Divide*, 87.
213 I. G. Barbour, *Religions and Science*, 167, 357.
214 S. L. Jaki, *Impassable Divide*, 90.

A principle that has no proof in science itself cannot therefore logically be used to describe the relationship between its subject and another in the same way as it is intended in science. Jaki therefore warns that we should not forget that the two sides are divided but the human mind, by the power of his reason has the ability to hover over the divide and do justice to two distinct domains lying on both sides of it. He thus writes: "To live in peace with that divide one needs a mind alive to its ability to hover over the divide, which is actually what the mind does when it performs even its most elementary functions."[215] This is man's ability to use philosophy stemming from wonder and the search for universal principles that underlie various and diverse realities, which alludes to the one and the many of the ancient philosophers.

If we are to talk of complementarity between science and religion, then we have to derive it from the two domains giving man the investigator and the believer a complete view of the universe not in the sense Bohr understood complementarity as an epistemological progress of the two sides. The difficulty raised by Jaki on the use of word complementarity has made it necessary for us to look at the roots of this term to understand it better and how it can be applied in the context of science and religion. The term 'complementarity' comes from the substantive complementary, which is derived from the Latin, in the composition between the preposition *'cum'* and the root *'ple'* which means, 'to fill'. This, in turn, looks like the transliteration of the Greek word *'symplerotikos'*, again with the preposition *'sym'* for '*con*', or 'with' in English and the stem '*plero*' to indicate that which is filled in, and it is full. The term therefore, as used by Bohr, means two elements filling in together to arrive at the completeness of a reality. He indicates the conditions for a "complete" knowledge of phenomena of wave properties and the properties of particles. Bohr complementarity principle states that a complete knowledge of phenomena on atomic dimensions requires a description of both wave and particle properties. Taken in this way, the complete knowledge cannot be arrived at if waves and particles of waves do not fill in each other.

215 Cf. *Ibidem.*, pp. 96–97.

The word complementarity therefore, cannot be used in same way when used in physics as when used to compare the relationship between science and religion. We have to first understand what is complemented by science and religion. The knowledge of science and religion complement, in the sense of filling in, the knowledge of the thinking being in order to have a holistic vision of realities in the universe. In the above case therefore, it does not mean that science and religion complete each other in the sense of science increasing the faith of the theologian and religion educating the scientist in the field of metaphysics and thus making science more scientific. It does not also mean that science and religion cannot be understood independently of each other as separate disciplines. The term is therefore, used by extension in philosophy to mean that science and religion complement the thinking and reasoning being, that is, the human person. This integration takes place in the rational capacity of man because the human reason is just only one. Complementarity in a philosophical sense makes the scientist and the believer achieve the unity between conceptually separate domains each contributing to the unrelenting desire of man to know natural things.

Both sides of the divide therefore have their feet on the ground parting from concrete realities and progressing with different methods, using the power of reason, to the acquisition of different types of knowledge that serves the same rational being who is satisfied not only by rationalizing, but also by examining realities and penetrating into the depths of what they present to the senses. Jaki argues:

> For both sides are real sides and therefore deal with the same reality though in two very different ways. Science is satisfied with the quantitative which reality reveals about itself and goes on from there to further sets of quantities. The philosopher or theologian focusses on the reality itself and cannot go much further from there except to keep reflecting on what it means to be a being.[216]

The first thing to note is that they are real sides of the same reality and not just a product of some fantasy. The methods of investigation separate the two sides. While one looks into things and how they are made and function, another looks into the same thing and why they are the way they are, why there is something at all and not nothing, and what they are destined for.

---

216 S. L Jaki, *The Impassable Divide*, 99–100.

This process can occur in the same person who is a believer and a scientist. The militant approach that one cannot be both faithful to religion and science is defeated by the fact of the individual being capable of doing both. Jaki stands as a good example of this reality. The quantitative touch of reality is an aid to the philosopher or theologian but not a source that can help him epistemologically to systematically transcend the physical realm into the metaphysical without creating much confusion and contradictions. The real source for both is concrete reality itself given by nature. According to Jaki, scientific knowledge would be an advantage for the philosopher or a theologian, because "it may keep him open to the importance of the quantitative aspects of things and processes, even though his method does not qualify him to discourse on the exact science."[217] A philosopher who has good understanding of science is better placed to raise deep questions concerning existence than one who has no formation in it at all. He is able to promote integral knowledge of the cosmos because of the quality of his mind that can distinguish two separate things. Jaki asserts that "the question about the relation of science and theology needs not only a clear mind, but also a clear conscience, which knows something about the sin of intellectual cowardice. It takes courage to say [...] that religion and science lie on two sides of a conceptually impassable divide."[218] According to Jaki, misuse of scientific language, illicit jump, whose characteristic is drawing religious conclusions from scientific data or, vice versa, forceful concordance, and virtual harmonisation can all be avoided if science and religion are kept in their respective conceptual domains. The appeal to conscience seems to go far as tranquilly facing and seeking the truth with intellectual honesty and openness to truth that lie beyond the tools in our hands. A philosopher, a theologian and a scientist cannot escape the noble intellectual duty to present the grounds on which he can safely speak of the strict totality of things or the universe writ large.[219] For Jaki, the ground is the concrete tangible reality presented to our senses from which each departs to arrive to greater knowledge of the universe.

217 *Ibidem.*, p. 100.
218 *Ibidem.*, p. 105.
219 Cf. *Ibidem.*, p. 107.

We can therefore conclude from Jaki's arguments that he sees a clear harmonic relationship between science and religion promoted by good philosophy as a principal requirement for a proper comprehension of the two conceptual domains. This will be even clearer in what we shall discuss in the second part of this chapter. This relationship finds its integration in man's mental capacity to hover over the two divides. This is because both science and religion seek to explain natural realities using reason and in the process have the need to appeal to philosophy to coherently present these facts or truth in a comprehensive and systematic way. We conclude this section saying that for Jaki, science and religion are conceptually separated and can be intellectually integrated owing to philosophy and its methods. In Jaki's own words:

> The Rock bottom rule is that truth is one, though, of course, with different aspects, simply because the human mind cannot reduce to one set of conceptualisations all those nuances. Within the set of quantitative conceptualizations of empirical data, the findings of science rule supreme, provided they are securely established. Whatever is outside that set is non-quantitative and is best called philosophy and even metaphysics.[220]

We would like therefore, to examine the role of philosophy in the relationship between science and religion using Jaki's cosmological arguments.

2.4 Jaki's Cosmological Arguments and the Role of Philosophy

According to Jaki, a critical look at science leads rationally to a good conclusion about faith not because science answers to faith, but because from scientific facts cosmological arguments are made more viable and clearer and from that clarity one can legitimately raise metaphysical questions that leads to faith. Such a move is not possible without the use of philosophy as a bridge between what science crudely presents and the question it raises that needs explanations that it cannot give. Jaki argues that if the universe is to give any indication of a true God, which for long has been understood as the cosmological proof, then the primacy of philosophy is indispensable. He argues thus:

---

220 Id., *Bible and Science*, 211–212.

> It has to start with a drill, so to speak of the mind so that it may see in any ordinary thing, however trivial, that specificity which reveals its contingency, namely, that it could have been something else. Having undergone that drill, the mind will reverberate with the readiness to use each and every thing as a springboard to something else and will find it most natural to use the universe too as a board that catapults one to mental heights where one recognizes the maker of the universe.[221]

As we continue to deal with the rationality of science and Christian faith, and specifically looking at the relationship between science and religion, we cannot but pay keen attention to the role philosophy plays in it. What we understand by the drill of the mind is to put reason to proper use that can see connections in what science is capable of establishing with measurement and experimentation. One cannot see a spring board if one does not see a logical connection in the things that can be seen and touched. What Jaki calls ordinary things are experiential realities that are the first level of abstraction in the process of knowing. Reason, whose practice is philosophy, therefore, becomes the bridge from which metaphysical questions are raised and concluded. When Jaki makes reference to philosophy, he always refers to a philosophy driven by truth and not mere sophistry. This is important because there are many forms of thinking that do not qualify as good philosophy and if at all they are philosophies, then they are not founded on the search of truth for the sake of the truth and thus a lame form of philosophy, not to talk of it as bad philosophy. For Jaki philosophy must be a true love for wisdom, "that love is not a matter of sentimentalism but a commitment which, when necessary, should be fiercely strong in defending its own rights."[222] Such a commitment does not admit of any form of mediocrity but an adherence and consistency in the search of the truth. It also calls for self-awareness of one's competence and limits. The cosmos therefore points to something beyond itself only for those who are capable of cultivating the required intellectual rigour to decipher its content. In the absence of a strong mental rigour Jaki asserts that "the mind will see in the cosmological argument a leap in the dark, a leap for which a faith largely devoid of intellectual components will be needed."[223] The role

221 Id., *God and the Cosmologists*, 75.
222 Id., *The Impassable Divide*, 108.
223 Id., *God and the Cosmologists*, 57.

of philosophy here is to identify causal connections from which logically it arrives to the conclusions that are implicit in nature. Scientific findings therefore have their best explanations in philosophy outside the realm of science itself. This is possible thanks to the philosophical method that deals with universal other than specific principles.

If we are to deal with the cosmos, we cannot but have to deal directly with the contingency of the Universe. Science over the years has been able to establish the fact that the cosmos is contingent, but that this contingency cannot be investigated into scientifically. Contingence is best understood in philosophy; that is, an existent is contingent when its existence depends on another, who could cause or not cause such an existence. It means a being does not exist necessarily. This implies therefore that beyond physical reality, lies a metaphysical one that science has neither method nor tool to investigate. Jaki writes that "such a contingency means the thorough dependence of a thing on another factor, without assuring a necessary existence to that thing [...]"[224] This means that such a thing, if we are to refer to the universe, could have not been. Jaki continues to write: "Even when cosmologists do not write esoteric equations about the universe, it exists. While those equations may marvellously account for all quantitative specificities of the universe, they suppose its existence instead of assuring it."[225] It follows that affirming existence is not the work of the cosmologist or the physicist because in attempting to do so they will be doing metaphysics. Physicists cannot do metaphysics because it is incompetent to do it. Respecting scientific boundary would demand that such a role be handed over to the philosopher. Jaki argues thus: "Wide as the net of physics may be spread, its catch consists only of the quantitative properties of things."[226] This is implicit of the fact that what net of science does not catch because of the nature of its net still exist and forms part of the deposits of the knowable. The fact that science does not grasp it does not make it less important. The quantitative properties of things and non-quantitative properties cannot be reduced into one in the same way because they are

224 *Ibidem.*, p. 94.
225 *Ibidem.*, p. 100.
226 *Ibidem.*, p. 106.

separate categories. The very condition for the possibility of physics is the fact that the universe exists as a condition for its operation.

Jaki continues to show the limits of science and therefore the need for philosophy using Kurt Gödel's incompleteness theorems, which do not only show, according to him, that no theory is perfect but also that science is greatly limited to fully comprehend even the most trivial of matters in the universe.[227] We would like to assert that the limits of science however does not imply the limit of knowledge but a springboard from which another type of investigation should take over in the spirit of acquiring an integral knowledge of what nature provides.

In the course of history, some scientists have endeavoured to explain creation for example using quantum mechanics or the principle of the conservation of matter. According to Stephen Barr, the attempt to explain the origin of the universe using science was motivated at least by two major factors. The first was the search for that initial singularity which he defines as "a special point where something breaks down, often because some quantity becomes infinite."[228] This issue was raised by Albert Einstein to refer to time zero that scientifically accounts for the initial singularity of the universe.[229] This however should not be taken as science proving the singular act of creation sustained by religion, because creation is not a mathematical affair and its assertions are not based on scientific formulations. But because such an achievement on the part of science seems gigantic, scientists and some theologians are tempted to fuse the two realms in some form of conformity. To arrive at the singularity theory, scientists make so many simplifications of very complex natural reality. They make a progress, too, by using series of approximations. Even if that single point is close to the creation story, no religious conclusion can be made or should be made from it. There would be no scientific ground for such a conclusion.

The second motivation of the scientists is to account for the initial conditions. This does not stem directly from science but seems, according to me, to come from science reading into religion. If things were not at one point, it means that they began to exist in time to the point in which we

---

227 Cf. *Ibidem.*, p. 113.
228 S. M. Barr, *Modern Physics and Ancient Faith*, 269.
229 *Ibidem.*

find ourselves now. The question that arises is that owing to the fact of our separation by time between now and before, how do we get to know the initial state where there was nothing? This raises two major philosophical questions about the origin of things, but also logically about their end. Religion would say things in the cosmos were created by God and they are going back to God. But then why should this be a problem to science? Barr argues that the philosophical motivation to extend science to the denial of religious account of the single initial stage as an action of God is purely "theophobia." Perhaps it is worthwhile to quote him directly here:

> One can think of the initial condition of the universe as the cause, in some sense, of all the later development that take place. But who sets those initial conditions? One obvious candidate is God, and indeed this is precisely the role that is assigned to God by 'deists,' he wound up the universe like a watch and let it go. Atheist would like the cosmic watch to be self-winding.[230]

The battle here is against the theist ground that the initial state can be accounted for by the free act of God creating all that is from nothing, for a purpose and sustaining it in being. Moderate scientists, taking a deist position want to assert that once God has finished his work, the universe as we know it by science runs its course without the need of God. This struggle has driven scientists to attempt to explain creation using quantum theory, derived analogically from the creation of quantum particles, details of which we would not like to enter into. Our interest is to show that philosophy can settle this wrangle between science and religion and promote harmonic relationship between them. In the midst of this struggle scientists mixed both science and philosophy ending up in creating great confusion. It is apparent that scientific conclusions cannot be used directly to draw religious conclusion without the use of philosophical arguments as a bridge. Along this line Jaki argues that measurability cannot be put forward as a condition for existence and that creation out of nothing must precede the existence of whatever quantum mechanics deals with. Creation, being an ontological cause, cannot be measured mechanically and cannot therefore be explained by experimental science. To Jaki, be it Hawking's quantum creation, or Alan Harvey Guth's inflammatory theory, or that there is a zero

230 *Ibidem,*. p. 271.

point beyond which origin (cause) of the universe should be obtained, the universe remains a creation of a free will of God. According to Jaki even zero is a specific point on the scale connecting positive and negative quantities. He poses a very important question: "Is zero a mere numerical idea equivalent to that nothing which in ontology is the very opposite of real existence?"[231] He argues thus:

> Creation remains a strictly philosophical as well as theological matter. It would, of course be useless to argue a philosophical point with those contemptuous of that basis of philosophy which is the definition of the meaning of basic words. When physicists – cosmologists claim that perhaps the reason that there is something instead of nothing is that [the] nothing is unstable, he plays an ugly game with words, man's greatest reassure.[232]

Perhaps the greatest challenge of our time pointed out here is the contemptuous attitude towards what does not fit within the framework of our thinking or what lies beyond it. Contempt for philosophy is actually contempt for a unifying factor that can put mere data into a coherent body of knowledge that benefits an integral view of reality. Words, therefore, cannot be used to cover up truths if we are not to fall in the same pit of the sophists and relativists. If God is replaced as the wilful author of the universe, another god is created and named chance and 'random-auto-creator' which have more conceptual problems than trying to understand the theist God. This is so because the theist God, at least in the Christian sense, is not a mere concept but a person whose interest and operation in creation does not expire. We can say that science is rational in so far as scientific findings are not extended to mean what they do not manifest in their quantitative form. Fallacies made by scientists are scientist's fallacies and not scientific fallacies. If scientific cosmology points to a zero point of the origin of the universe, beyond which there must be a single force to realize something out of zero, it is because it is capable of rational procedure but incapable to be the answer to the explanation of the point beyond which it points. The initial stage zero of science is not the same with the nothing held by religion as the point of creation. This is because time zero already presupposes the existence of time and not nothing. This time is expressed mathematically as

231 S. L. Jaki, *God and the Cosmologists*, 142.
232 *Ibidem.*, p. 145.

zero making it something. Religion on the other hand sustains nothing as the initial condition which presupposes no time at all. The initial stage is not just a vacuum that God filled up with creation, not even zero but nothing. Science can therefore begin only from something in time from which experimentation is possible in terms of conditions before and after. Science cannot, therefore, speculate about time before time, except by theoretical extension from the present time, which is actually the conscious time after. To bridge the gap, science needs philosophy to intervene with concepts like absolute time and absolute space that rationally make sense. Cosmologists should have no obligation to explain the why of what its study reveals, or the initial conditions before existence, because that will demand an ontological explanation that reason puts beyond his tools and methods, and the same reason passes the button to the philosopher and the theologian.

2.4.1 Science, Chance and Philosophy

Jaki delves into the ancient problem of chance that has occupied philosophers throughout history and modern science has taken as either substitute to the causal explanation of the universe against creation by God or extremely as the ultimate cause of all things. The article of Lorella Congiunti "*La collocazione epistemological del "caso": all'incrocio dei saperi*" treats well the various usage of the term 'chance' in various disciplines.[233] In her article Congiunti identifies two philosophical problems raised by the use of the term chance to account for the origin of the universe by Hawkins. According to Congiunti, chance cannot account for the beginning because the absolute beginning cannot be known purely rationally because rational knowledge cannot get outside time in order to understand the first instance. Secondly, saying the universe is self-created is a contradiction in terms because what is created cannot create itself.[234] These two philosophical problems will return frequently as we discuss the rationality of science and Christian faith. Right from Plato chance had no place as the ultimate cause because it goes against purpose which is manifest in the universe.

233 Cf. L. Congiunti, "La collocazione epistemologica del "caso": all'incrocio dei saperi", in *Urbaniana University Journal*, 1 (2015), 109–140.
234 Cf. *Ibidem.*, pp. 121–122.

Aristotelian causality gives no space to chance even when it falls into pantheism.[235] According to Congiunti, in Aristotle, chance is related to a series of causes that may have not been intended as an initial plan and therefore and intersection of causes qualified as accidental causes. Chance therefore in Aristotle calls for analysis of causes.[236] It is clear that chance cannot be the ultimate cause and cannot be used to account for the origin of the universe. Jaki asserts that "even more emphatically, chance as a material cause is rejected by Aquinas. He, of course, admits the coming into being out of nothing but, in the Christian credal perspective this process can arise only through the creative power of God."[237] For Aquinas, God and only God can create out of nothing.[238] In this is implicit the idea that no creature can be a creator and that there must only be one creator. This creator should also be outside creation and not part of it. Aquinas uses the context of chance as an accidental cause, taken from Aristotle to demonstrate the divine governance of the universe. God did not create and go on holiday; he continues to govern his creation through providence. We shall not enter into the discussion of providence since it will be tilting us towards a theological discussion that I do not intend to do in this exposition. But what happened that our Christian modern world seems to go back to chance? According to Jaki, "Christian understanding of chance became less and less a part of western intellectual consciousness as the latter moved from renaissance through enlightenment times towards the present, markedly post Christian age."[239] Jaki discusses how various philosophers had different definition of what they considered as chance showing their weak points as well. For David Hume, "chance is only our ignorance of real causes", while Montesquieu and Voltaire held that "chance can only be an unknown cause of a known effect." He fell into extreme determinism. For Claude Adrien Helvetius, "I understand by chance the unknown chain of causes capable

235 Cf. S. L Jaki. *God and the Cosmos*, 149–151.

236 Cf. L. Congiunti, "La collocazione epistemologica del "caso": all'incrocio dei saperi", 128–129.

237 S. L. Jaki, *God and the Cosmologists*, 151.

238 Aquinas, *Summa Theologiae*, I q. 45, a. 5. (Translated by Baumgarth P. William and REGAN J. Richard, University of Scranton Press, London 1994).

239 S. L. Jaki *God and the Cosmologists*, 153.

of producing such and such effects." And for Thomas Henry Huxley "do the [Darwinists] believe", he asked, "that anything in this universe happens without reason or without a cause"?, and finally Henri Poincare: "chance is the measure of our ignorance."[240] In all these attempts to explain what chance is there are three strong desires or attitudes that preoccupied the philosophers and historians. Firstly, that the mind will be able by the use of science to establish the cause of everything and it is just a matter of time. This is an attitude that seeks strictly a scientific explanation of all things as opposed to religious believes. Secondly, that operations in the universe are predetermined by nature and they will continue to follow the internal law in them. An attempt to defend eternity of the universe. Finally, that the exact cause of things may be forever unknown to the human mind, which is the sceptic approach to the understanding of the universe. The quest for the knowledge of why there is something and not nothing cannot be answered by speculations or reduction to what is known, scepticism or experimentation in the laboratories; it requires a major openness to a good philosophy that analyses universal principles that underlie existence. It is only philosophy that can explain being in a way that can be understood using metaphysics which par excellence is the philosophy of being. Natural reality must first be before it is something.

Jaki argues that in the 20th century however, chance became something that was likened to a cushion. He asserts, "chance became a philosophical pillow in all scientific history in the 20th century."[157] He quotes Werner Karl Heisenberg to bring out an argument that should never be taken for granted: "The physicist must, however, postulate in his science that he is studying a world which he himself has not made, and which would be present, essentially unchanged, if he were not there."[241] Even if Heisenberg did not arrive at the point of acknowledging God as the creator of all things and author of order, he raises a very important point that scientists deal with what is given, which continues in being whether they will ever know it or not. Existence does not depend on our knowing it. If the universe is therefore a product of chance, then it must be produced by chaos and

240 *Ibidem.*, p. 154.
241 *Ibidem.*, p. 161.

if it is a product of chaos, then implicitly science would not be possible and whatever scientists would be doing would be irrational. It is therefore important to take into consideration the arguments of Congiunti that the term chance is used in various disciplines from divergent angles, methods and intentions and therefore, the same meaning cannot be applied from one discipline to another without careful mediation.[242] Terms should therefore be defined and proper language used for particular fields of studies. If there arises any need to extend the use of a term from one discipline to another, philosophical-contextual mediation would be an indispensable requirement. Talking of God as the creator would therefore give no room for the application of chance as a cause in the same way as God is. Chance as an accidental cause therefore, inheres in the ultimate cause of all things, who is God himself.

There is therefore great need for the training of the human mind so that he is propelled towards an upward mobility, towards God, "the only being beyond the universe of specific."[243] Jaki warns that the move towards God should not be an abstract move if it is to be safe, it must have a firm foundation in the universe. "The move rather consists in sensing the pulse of cosmic contingency, the relentless pointing of the universe beyond itself."[244] The universe is only a signpost to its author because it is contingent, and it makes sense since all that there is in the universe is not God. Jaki argues that if the universe is to point or lead to God, then the human mind must first be purged off half-truths. In his words he says:

> The cosmological argument will appear as a delusion unless those fallacies and half-truths, which the modern world mass-produces about the universe, are seen for what they truly are. It is precisely because the modern intellectual atmosphere is polluted to suffocation with disastrous ideas about the universe that any vote cast on behalf of the cosmological arguments may appear a sheer defiance of all accepted standards of intellectual respectability.[245]

---

242 L. CONGIUNTI, "La collocazione epistemologica del "caso": all'incrocio dei saperi", 126.
243 S. L. JAKI *God and the Cosmologists*, 223.
244 *Ibidem.*
245 *Ibidem.*

The desire for God is an informed disposition that cannot be shut down. But in a world full of explanations about God, there is great need for an intellectual formation to discern good explanations from bad ones. Moving from the universe should be based on the truths that are concrete. Jaki calls for firmness if the cosmological argument should make sense as a proof for the existence of God. He argues that:

> The effectiveness of the cosmological argument depends on whether one can take for certain the elementary everyday realities. To take something for certain is not an act of faith, but an act of plain intellectual recognition of its reality. Any proof relating to reality depends on one's confidence that one knows things, but the intellectual act does not thereby become a matter of emotions with resulting loss of its objective certainty.[246]

The above quotation answers many puzzles of our modern time. Jaki, from a realist point of view, advocates for a recognition *simpliciter* of the objects in everyday life as real and if they are concrete and not abstract realities they object to our senses and demand of us an intellectual response.[247] The first rational step in understanding an object of reality is the elimination of illusions and the acknowledgement that they exist the way they do. This in itself is a prove of our rationality and a step towards some objective truths. Concrete reality does not therefore depend on the perceiver; it is true that I am writing right now, the perception of another does not make my writing false. As much as the perception of another will not make false the fact that you are reading this sentence right now. Ignorance of a third party that you are reading right now does not make your reading any less true. If this is applied to concrete daily realities, then science and faith are grounded on the universe and what it contains that forms an integral part of human cognition. According to Jaki, our knowledge of reality is often accompanied by uncertainties that can only be assured by the truth.

> This is why truth is given its greatest source when its certainty is held high unconditionally […] The cosmological argument too, will exude a vigorous demonstrative force only if no hesitation is allowed about the fact that man can truly know,

---

246 *Ibidem.*, p. 232.

247 Cf. A. Giostra, *Stanley Jaki: Science and Faith in a Realist Perspective*, IF Press, Rome 2019. Cf. P. Haffner, *Creation and Scientific Creativity: A Study in the Thoughts of S. L. Jaki,* Gracewing, Leominster, UK, 2009, 171–185.

> that to know is to know something and that whenever he knows, he perceives the universal, the whole, in the particular.[248]

The ability to know universals in the particular will be for man the sure foundation for which he can transcend physical reality. Jaki responds with a realist mind to the disaster of relativism in the modern world. If we know something, then that knowledge of something must be real and objective because it is based on real things as opposed to fantasy. This is the first step towards a good philosophy that has roots in real intellectual encounter with objects. Jaki's argument is not just about having a philosophy, but more so having a sound philosophy, that begins from intellectual honesty, and concrete reality if science is to have a realist future. Jaki argues that "once the physicist had departed from that philosophical middle road in which expressions like "to be" and "to know" can have unambiguous meaning, he could not remain immune to the inner logic that was to toss him from one unscientific extreme to another."[249] By referring to philosophy as "the middle road" Jaki explicitly underscores the bridging role of philosophy between science and religion. The only mediator capable of making the two ends reachable by the human mind is philosophy with its unifying method. The fundamental duty of science is to expose reality into some understanding and according to Jaki "one's knowing is meaningless unless one knows something, that is, unless one's knowledge touches on reality."[250] This is at all times accompanied by reality itself and not just merely how one can understand reality. At the base of this argument is the criticism of Kant and his systems that Jaki out rightly dismisses on the ground that Kant was not a scientist and yet he claimed so much in the name of science.

He concludes: "the spokesman of the cosmological argument will therefore know that only by intellectual firmness will serve the purpose of that argument and those who want to profit by it."[251] It is implicit therefore that shallow arguments for or against the cosmological argument is a manifestation of a lack of intellectual rigour and philosophical foundation. Such a manifestation of the lack of intellectual firmness is even worse when done

---

248 S. L. Jaki, *God and the Cosmologists*, 232.
249 *Ibidem.,* p. 207.
250 *Ibidem.,* p. 212.
251 Id., *God and the Cosmologists*, 232.

by one who attempts to defend the cosmological argument in the name of the Christian faith. To argue that we cannot be certain of the universals as some positivists argue is actually to deny man his basic mental ability of abstracting universals for which that same negation would not be possible. "To discover the universe means journeying beyond the universe."[252] This journeying cannot be anything less than an intellectual and reasonable journey. The scholastics like Thomas Aquinas have long demonstrated man's ability of grasping the universals using reason and intellect, a fact that cannot be contested if man is to grasp anything that makes the knowledge of physical reality possible.

Jaki argues that "in physics any particular datum can be traced only to another such quantitative datum. Much less can any physical mechanism, any physical force or process, no matter how fundamental, be identified with God's there, although He is most imminently present everywhere."[253] Here lies an argument for the rationality of science when practiced well. It always produces results that are logically within its competence moving from quantitative datum of physical substance to more quantitative datum of the same. That is what science is and what it is best at, without any extension of its competence. Jaki argues:

> God has no place in science, nor any denial of Him or any scoffing at Him. Nor is it good scientific taste to take science for a cover for scurrilousness about what is the ultimate in being and intelligibility science should also be distinguished from any broader philosophical, but alone theological interpretation of it.[254]

Reason alone should be shy to merge scientific knowledge with metaphysical one because they lie at different levels of abstractions. This should not be seen as negating other sources of knowledge about our common home, but an openness to that fact that the universe offers so much more than we can fully comprehend, even when all forms of knowledge are put together. We recall therefore that even if a scientist and a philosopher are in search of the response to the same question of the why and what of things, they both seek and arrive to different answers permitted by their methods and in context of their investigations. While the scientist looks for his answers

252 *Ibidem.*, p. 236.
253 *Ibidem.*, p. 264.
254 *Ibidem.*

in quantities, the philosopher looks for his answers in what lies beyond quantitative data that answers his puzzle. Science can have as much data as it is capable of investigating, but it needs philosophy to draw out universal principles from the totality of the things that is and finally produce an integral vision of the universe that makes science and religion relevant.

2.4.2 The Problem of Evolution, Religion and Philosophy

One of the controversial issues in the history of science that has provoked questions about the relationship between science and religion is the evolution theory developed by Charles Darwin. It is worth treating the evolution problem on its own as it raises many philosophical and scientific questions regarding religion. The conclusion that each species had not been independently created, but had descended, like varieties, from other species, changed the vision of the world and its author. The major challenges to the Christian understanding revolve around; the authority of the Bible about creation, the argument from design that leads to a creator God, the primacy of humans over other creatures and the moral and ethical foundations of behaviours.[255] These are serious questions that we ought to deal with as we try to understand creation both in the manner of science and in religious terms.

The term 'evolution' is etymologically derived from the Latin *evolutio*, which means a gradual development or a process of transformation. Interpreted in this manner, for example, means that man as man now is a product of millions of changes with time. The point of conflict is however, not the gradual process of development but when scholars begin to interpret evolution in relation to the Christian creation story. Evolutionists assert that living beings were not created by any other power but that they evolved from matter. Jaki has endeavoured to respond to this challenge by publishing several articles and books notable of which is his book, *Cosmos and Creator* (1980), in which he treats in the third chapter the 'Dogma of

---

255 Historical and contextual readings about evolution can also be found in I. G. BARBOUR, *Religion and Science: Historical and Contemporary views*, 49–72, 222–245. Confer also S. BARR, *Modern Physics and Ancient Faith*, 72–75. Good philosophical arguments based on Aristotelian four causes can be found in R. J. CONNELL, *Nature's Causes*, Peter Lang, New York, 143–181.

Creation', discussing science and creation.[256] Another publication in which he deals with Darwin and Darwinism is his *The Saviour of Science* (2000), in which he discusses 'Darwinists in the dock of Metaphysic.'[257] *Bible and Science* (1996), in which he discusses how the Bible has been mistakenly taken for a science text book that contrast with experimental facts. He argues in it that both scientific facts and the revelation in the Bible contribute greatly in their different ways to the holistic understanding of the universe and man's place in it. There are several of his articles on the interpretation of the book of Genesis and concerns about the evolution theory that can be explored as areas for further research.

Darwin was neither a philosopher nor a theologian, but a biologist whose theory has been extended to philosophy and theology with the consequences that they bear. It can be noted that even in Darwin's evolutionism there are many assumptions that are not proven; the projection into the past is unproven due to the fact that time and the shortness of our lives separates us from it. Evolutionists cannot therefore accuse religion for holding believes that are unproven, because they actually have the authority of revelation unlike the assumptions of the evolution theory. It is for the above reason that we need to explore the dialogue between faith and evolution in science as presented by Stanley Jaki taking the line of philosophical arguments.

According to Jaki, there is a thin line that Darwinists and Darwin left between the theory of evolution and ideology. This is because using evolution theory as an explanation of the origin of species is purely ideology mixed with science. This creates confusion in the understanding of act and potence.[258] Jaki's argument here goes back to what science should be. It is possible that species could have evolved in some way, but it is not actual that they evolved until there are scientific proves for it. Potency therefore does not guarantee actuality especially when we are dealing with speculative theories. Jaki argues that:

> overwhelmed as they were and still are by Darwin's appeal to the imagination, his admirers think that this difference between "could be" and "having been done" can be taken lightly. In addition, they are motivated by ideological considerations

256 C.f. S. Jaki, *Cosmos and Creator*, 56–87.
257 C.f. Id., *The Saviour of Science*, 221–243.
258 Cf. Id., *Lectures in the Vatican Gardens*, 170.

> that vary from crude materialism to misguided theism, the result is a huge imbalance between what is proved and what is assumed to have been proved.[259]

If ideologies are sets of assertions that are taken wholesomely without questioning its validity, then evolution theory promoted by Darwinists are extended beyond its scope to mean that evolution can answer the question about the beginning of life and diversity of species. Darwin himself has not been able to account for a transition of one species to another by observations, which is a basic requirement for science to be science. When we talk of evolution in a philosophical and more serious way, we ought to take seriously the relation of cause–effects by which we can understand well the creation of the universe. To emphasize even more that there was more ideology to do with evolution than science, Jaki quotes Ernst Chain:

> Evolution by chance is a [mean] hypothesis based on no evidence and [is] irreconcilable with facts. [. . .] Evolutionary theories are a gross oversimplification of an immensely complex and intricate mass of facts, and it amazes me that they are swallowed so uncritically and readily on for such a long time by so many scientists without a murmur of protest.[260]

Evolution theory moved from a hypothesis to a theory that in itself has no proof. To make matters even more difficult, the closeness of humans to apes has never made one side to cross into the other "That human chromosomes differ but slightly from those of higher apes only increases the problem of why humans, and they alone, think, speak and have science as well, all super-gigantic differences in respect to apes."[261] What evolution can rightly account for are mere development and changes within the same species due to adaptation to environment which is a potentiality in them. What Darwin and Darwinists mixed up was science and ideology, what Jaki calls mixing carful "observation of facts with loose reasoning."[262] This is definitely a departure from both philosophy and science, or at least reasoned science if not exact science. Transient forms that were largely accounted for by potentiality was under looked by Darwin and philosophically taken as act and thus producing many forms that most of them left no trace at

---

259 Cf. *Ibidem.*
260 Id., *God and the Cosmologists*, 175.
261 Id., *Lectures in the Vatican Gardens*, 177.
262 *Ibidem.*

all because they were just a stage in the life of the actual species.[263] The problem with this is that it is difficult to distinguish the point from which evolution as origin and evolution as transformation begins. Being more of ideology than science, Darwin and Darwinist mix these moments as if they mean the same thing. This calls for some consistency and intellectual honesty on the part of the evolutionists. Jaki continues to argue that "for if the transformation of one species into another came in a sudden burst, either the transitional form had to be still visible, though not piled upon one another in the relatively thin strata [...]"[264] The argument of the lack of evidence (observable) transformation of one substantial form to another makes the evolution theory incapable of explaining the ultimate origin of species in the universe because something must first be in order to evolve. Evolution science cannot account for the origin of that matter from which evolution is possible.

One wonders why the evolution theory and Darwinists' promotion of it should create issues with faith or religion. Jaki gives a very sharp judgement and reason for it: "Darwinists all too often acted as if they formed a church of anti-religion."[265] This accrues out of the desire to give an alternative explanation of the origin using irrational science mixed with ideology. Every Darwinists pounded on the ideology that their theory gave a comprehensive view of life and existence. Titles like *Religion without Revelation* of Julia Huxley speak for themselves.[266] It goes back to the militant confrontation with religion.

Jaki discusses glaring contradictions in the evolution theory brought by the behaviour of some species that neither obey natural selection, nor survival for the fittest nor any evolutionary change. Some of the examples he gives include the behaviour of the water spider, botanically called *argyroneta natans*, which basically lives under water and makes in an extraordinary way its nest at the base of the water with a connection for aeration on the surface of the water. The trap door spider that makes its nest in a

263 Cf. Id., *The Saviour of Science*, 155.
264 Id., *The Saviour of Science*, 155.
265 *Ibidem.*, p. 159.
266 *Ibidem.*, p. 161.

very intelligent way and defeats the argument of natural selection.[267] These examples demonstrate the impossibility of proving experimentally Darwin's theory of natural selection. If natural selection were really the mode in which species evolved, we would have the world full of perfect species and if adaption were a radical change in the species, humans would have the best pair of eyes and a very big head adapted to the functions they perform. It seems clear that the existence of the human person cannot be accounted for purely from a materialistic point of view as Darwin and Darwinists would wish to present it.

About the evolutionists argument that man is purely matter and that everything in the universe can be accounted for from a materialistic point of view alone. Jaki points out elements of human existence that already contradict that assumption. The intellectual operations of man cannot at any rate be accounted for in a purely materialistic way. Jaki asserts that "mental view goes infinitely beyond the physical view and conjures up what only the eye of the mind shall ever see"[268] Man's place in the universe, therefore, cannot be confined to material existence alone because his act of thinking alone is already immaterial. If evolution is to be understood as a purely materialistic explanation of the existence of all that is in the descent, then it cannot account for the minds immaterialist function and therefore reduces man's mind to mere matter going against itself. Reason dictates by its very nature and function that physical realities are not an end in themselves. Darwinists, therefore, would simply fail to account for the existence of the universe which is an abstraction from order and harmony in the cosmos.[269] The matter from which evolution is possible is evidently not disordered and has within it some natural laws it follows that makes a purely materialistic account an incomplete vision of reality. Jaki continues to assert that there is one strong argument against Darwinist's material evolution and anti-religion ideology "that there is a metaphysical vision upon which scientists as well as non-scientists fall back on each time they utter a phrase, however trivial, let alone when they offer daring generalizations."[270]

267 *Ibidem.,* pp. 149–151.
268 *Ibidem.,* p. 166.
269 Cf. *Ibidem.*, pp. 167–169.
270 *Ibidem.*, p. 167.

This is a dictate of good reason because the human mind has the capacity to go infinitely beyond the physical view of reality using the "eye" of the mind. It is the mind's function to grasp the non-material complexity and principles that underlie physical things.

The assertion that "the enormous increase in the exactness of physics in the 20th century called for a use of the brain that could not have been more brainy" goes a long way to underscore the importance of employing reason more rigorously for a comprehensive and integral world view.[271] This is what makes science a reasonable endeavour. In praise of good use of reason Jaki asserts: "An equal respect for both matter and mind should appear more and more the key to success as well as survival."[272] The core of our argument here is the unicity of the intellect that purify itself of contradictions that goes against reason and cognition. Stressing any extreme would jeopardize a balanced vision worthy of a sober mind. This kind of balance was practiced by the scholastics in the medieval period from which according to Jaki exact science was born. St. Thomas remains an unchallenged champion in the promotion of the truth of the immortality of the soul which is actually what defines the mind.[273] So to say that there is something beyond material existence that does not obey the adaptation or the evolution theory.

We can make the following remarks as far as evolution and Darwinism is concerned. The first is that the theory cannot account for creation from nothing. Evolution will always have to begin from something and that something cannot be accounted for within the same theory of evolution. Life is therefore more than its material components. Secondly, the evolution theory cannot account for the transition of one specie from another, that apes were not apes from day one apart from stating it theoretically, and therefore cannot account for the creation of individual persons, held by Christian faith as an intention of a creator called God. Purpose that can be inferred purely from the behaviour of physical things cannot be denied without contradiction by the theory of evolution. Thirdly, the creation referred to in the evolution theory has completely a different notion of what

---

271 *Ibidem.*
272 *Ibidem.*, p. 171.
273 Cf. *Ibidem.*, p. 174.

Christians would know of creation. There is an illicit use of the language of theology in a pseudo-science of evolution without any logical justification. Creation is not a mere transformation from one state of being to another, nor is it making from what is but it is an act of making what had no existence to exist. The Christian understanding of creation is strongly linked to the beginning, understood with the beginning alludes to nothing previously in existence.[274]

According to Richard J. Connell, even the terms 'evolution' and 'to evolve' should be properly understood. When we say that humans evolved, do we mean from a more primitive form to the current form, or do we mean from a particular form of a species to another? We have already seen that there is no observable evidence of evolution of one species to another, if not only a transient formal change like in metamorphosis.[275] It can be argued that the change in properties of a substance does not constitute substantial change that renders it radically different. When the change, is not a substantial change then we can say an organism has evolved over time or in a twinkle. But if the change is irreducible, without substantial identity to its original form then we may have to assume that evolution has taken place. That genetic mutation can occur is indisputable and biological but that the cause of species can be attributed to chance is not a biological question but, a philosophical one that touches religious and metaphysical issues.[276]

Evolution, therefore, raises very difficult questions as far as cause is concerned. Going back to Newton's law of motion, and applying it to cause, it can be argued that the cause of an agent must be greater than its effects, and thus causes must be proportional to the effects, and whatever comes to be must come to be from an agent that must in some way contain the effects to be produced. "To say that evolution is responsible for the origin of species is to say that it is the cause of the natures of species."[277] What evolution describes, however, is a mere rearrangement in the life of the same species in order to adapt to environment and mode of its living. These rearrangements, however, complex they may be, remain instrumental causes

---

274 Cf. Id., *The Cosmos and Creator*, 69.
275 R. J. Connell, *Nature's Causes*, 144.
276 *Ibidem.*, pp. 143–181.
277 *Ibidem.*, p. 162.

that are driven by the nature of the species itself, whose cause should be outside the species. The active cause must contain the energy proportioned to the effect. The nature of the species and its origin cannot be a random chemical combination. To say that the organic compound chemically combines to produce different organisms, is to say that an antecedent substance that "does not have an operational capacity grounded within it can be the agent cause of an offspring that does have the capacity."[278] This would be a contradiction of the gradual process proposed by evolution theory. It would be more plausible to say that a posterior species developed or evolved from those that were prior and thus not constituting origin of species but origin of accidental changes in the species. They are accidental changes because they inhere in the substantial form that remains unchanged. When the jawbones and teeth of the modern man become less robust because of the mastication of soft food, and therefore has to regularly visit a dentist, it does not mean that man has substantially changed. The change in his jawbones and the quality of his teeth are all accidental changes that does not alter substantially the nature of man.

Because facts are material components of science, in the spirit of intellectual honesty, we must concede that species came into existence successively and that present species evolved from prior ones. We must also concede that prior species could not have been the authors of themselves. This is important because the doctrine autocreation destroys itself by combining primary causality into one with secondary causality. Such a contradiction can only come from a mind that is not well cultivated both in science and philosophy.

> We infer not postulate or hypothesize the existence of another, principal causal action that is proportioned to the actual effect, the nature of the new species. And since the principal causal action is not that of another organism, of a material agent discernible directly through experience, we must also conclude that the principal agent is extrinsic to the realm of nature.[279]

This we call the universal principle which is responsible for the existence of a species without itself being species. This from a faith point of view is God but not in the sense that God creates within the framework of evolution,

278 *Ibidem.*, p. 164.
279 *Ibidem.*, p. 165.

but in the sense that he is the ultimate cause that is outside the things that he causes. We infer this and not hypothesize nor postulate because we depart from phenomena from which reason is able to transcend the realm of the material realities. The agent that is responsible for the formal causes of biological species must be outside species or extrinsic to it. Since our inference cannot be made without observation of natural phenomena, it can be argued that scientists cannot proceed without parting from the same point and if both are kept within good reason, they both will ask metaphysical questions whose answers can only be comprehensively given by the philosopher and the theologian. I would like to conclude this section by giving a definition of evolution sustained by Jaki that when taken strictly does not contradict creation as the work of God, Jaki says, "by evolution I simply meant that the powers of matter are wholly sufficient to account for any material transformation of species."[280] This definition does not deny matter its creator who endowed it from the beginning with the power of becoming. Evolution as a potentiality for adaptation and transformation of organism does not go against Christian doctrine. It would not therefore be out of place when Jaki asserts that "the only solid ground for holding evolution is belief in the createdness of the universe, and therefore in the strict interconnectedness of all its parts, a feature demanded by the infinite rationality of the creator."[281] What should be clear is that evolution theory cannot account for the origin of the universe and the things in it.

2.5 Concluding Remarks

In this chapter we have seen the four models of science-religion relationship as championed by Ian Barbour, in which he proposes the conflict, independent, dialogue and integration models. These models have characterized the relationship between science and religion in the course of history. Having established the general grounds of the relationship between science and religion, we went ahead to discuss particularly Jaki's understanding of the same relationship trying to collocate his thoughts in the four models proposed by Barbour. We exposed that for Jaki, there is no conflict between

280 S. L. Jaki, *Mind's Matter: An Intellectual Autobiography,* 98.
281 Id., *Bible and Science*, 146.

science and religion as long as one cares to strictly and clearly define the two. Jaki argues that what many other scholars term conflicts have been moments of clashes in history that were resolved. We however noted that there is a thin dividing line between a clash and conflict and that one leads into another. We saw that Jaki, by arguing for the inexistence of conflict between science and religion also eliminates the possibility of situating his thoughts in the independence model. We then went ahead to discuss Jaki's conceptual distinction that science and religion lie within two permanently distinct conceptual domains and the best thing the mind can do is to hover over the two in order to decipher the content of the universe. He argues that any attempt to mix the two or develop a universal theory for the existence of the universe would produce nothing other than confusion. In his famous and very original analogical assertion he says, what God has separated in different conceptual domains, no man should put together. He argues that reason has the capacity to ponder both the facts of science and the truths of religion without any contradiction whatsoever. We discussed how harmonic relationship can be fostered by the mind sharpened by good philosophy. The *aut* advocacy therefore is a door to reductionism that can be avoided. We noted that Jaki sees a harmonic relationship between science and religion and a possibility of an intellectual integration by use of philosophy. Being a scientist, priest, philosopher and theologian, he has demonstrated the fact that one can be simultaneously a scientist and a religious. In this chapter we also dealt with Jaki's cosmological arguments which is one of his direct philosophical works in which he demonstrates that the universe is not just a mental concept, but a concrete reality, because it is abstracted from real existence acting in unity in their diverse natures. This we did in an attempt to show the primacy of philosophy and the role it plays as a bridge between science and religion. We discussed the various philosophical and scientific views that show the rationality of science and the Christian faith. We also establish that both science and religion are capable of asking metaphysical questions but they both need philosophy as a bridge with its tools concerned with the universal principles. We discussed creation of the universe and cosmology and established that the biblical account of the creation cannot be taken as a record in a text book. We discussed the challenges of evolution theory developed by Charles Darwin and Darwinists and demonstrated that evolution has no proof of the development of species from one kind to another, but accounts for the development within specific species over time and above all that evolution is a theory very much mixed with ideologies that makes it [evolution] fall outside good science. We have seen that, unfortunately, these ideologies tend to be militant against religion and faith in God the creator. We saw that science is mainly concerned with facts that are fluid and can change over time with more accuracy, while religion and faith deal with truth which has an ontological dimension that falls beyond the competence of the scientific method. When science and religion are kept within their respective camps, with respect to the limits of their methods, they can go a long way in developing a complete vision and understanding of the universe and what it contains with the help of philosophy.

Chapter Three The Unity of Reason

3.1 Introduction

Jaki spent the major part of life reading and writing the history of science: the historical approach that demonstrates that science was born of Christianity is a plausible endeavour that cannot be denied or ignored. That is why I would like to dedicate this chapter to the examination of Jaki's arguments about the philosophical rationale that should accompany every scientist, theologian and philosopher who wants to talk about the relationships between science and faith. As already stated, the vocation of science should not be the deconstruction of the Christian faith, but a dedication to the search of knowledge that serves the common good, and in the same way Christian faith should not be the denial of scientific revolution, but appreciation of what science is capable of doing in rendering a good life to the human person by leading him away from superstitions. We have arrived at the concept of 'distinguishing to unite' faith and science, by the power of reason.

Owing to the fact that I am dealing with philosophy, I would like to demonstrate its role as a rational discourse, in the unity of knowledge and the holistic approach to reality that man is capable of. It is true that we live in an ambient of the fragmentation of knowledge that puts every serious scholar in an intellectual crisis and yet human reason dictates of its very nature an intrinsic unity if it is not to contradict itself.[282] I would like to assert in this chapter that the fragmentation of knowledge militates against the intrinsic nature of reason that seeks to unify in itself all that is knowable to have a holistic and stable world view. In this chapter I will attempt to demonstrate how our modern time needs a resolute return to the use of rigorous reason and fidelity to the search of truth, what Jaki calls most of the time the true love for wisdom, a love that goes beyond sentiments and passion. Ashley, in advocating for the cultivation of wisdom writes, "therefore, however original we may be, we all need some kind of

---

282 Cf. B. M. Ashley, *The Way Towards Wisdom,* University of Notre Dame Press, Notre Dame (IN) 2006, EPUB e-book.

wisdom, our own and that derived from others wiser than we are, that takes account of all information accessible to us and unifies it in a useable way."[283] This is a call to being comprehensive, critical, intellectually honest and asking relevant questions even when we may not be able to provide or find all the answers. I will also examine rationality and natural theology as a philosophical engagement. I will discuss the role of ethics in science and why Jaki thinks that scientist must have ethical considerations when they set out to do their experiments. We shall elaborate on the role of faith in science and see the development of science aided by mathematics. We shall then point out some misconceptions that are posited in the name of science about God and faith, and conclude the chapter with some discussions on faith and reason.

This chapter therefore, will centre around the premise that human reason is one, and since it is one, it dictates upon the thinking being its singular unity with multiple potentiality to grasp diverse concepts, be it scientific, religious, philosophical, and theological.

3.2 The Need for Rigorous Use of Reason

In his early works Jaki presented historical facts as they are because he was interested mostly in the reconstruction of the history of science in relation to Christianity. One of the objectives of this book is to reconstruct Jaki's philosophical thoughts, I would like to discuss his arguments that go beyond mere presentation of facts or appeal to Biblical beliefs that can be valid even for non-Christian audiences. The fact that Christianity provided a conducive atmosphere for the viable birth of science is not true simply because one is a Christian but it should be valid for the rational evidences advanced in its defence. This is not intended to mean that Christianity as a religion is merely better than any other religion, but simply that her worldview made it possible for science to be born as it is argued by Jaki.

Since truth cannot contradict truth if it is true, we can assert that reason cannot contradict reason in the same person. If faith is rational and science is a product of intended reasoned investigations, then, they both contribute in their own ways to the life of the human person and probably his happiness.

---

283 *Ibidem.*

The scientist therefore moves illicitly from the realm of observations and measurements to the metaphysical realm and yet he cannot avoid the metaphysical question if he is to be faithful to the demands of the unity of reason. As much as the scientist is capable of raising metaphysical questions from the evidences of his investigations, he has no strict obligation to give answers to them, if he does so, he should be aware that he is crossing the boundary of science into another discipline that requires another method and competence. If a scientist ventures into answering metaphysical and philosophical questions, he should be prepared to take into consideration a complete philosophical tradition and not merely a small portion of it. It seems safer therefore, to cultivate some intellectual honesty and remain within one's competence. The methods of science alone cannot arrive to God, but can point towards God when carefully reasoned out. Stacy Trasancos puts the need for this dependence very vividly, "what counts in the study of history is where the instruction leads the next generations and modern science, mature and independent as this child of the human intellect may seem – is desperately in need of its mother."[284] The advancement of science leads every generation to higher and higher comprehension of the deposits of nature. This can be a blessing but can also be a problem if the modern man forgets the foundation of his science, that should be the aid to give it a proper direction that serves at all times the common good.

Jaki reasoned that neither philosophy nor theology can be termed science in the same way physics, can and that is agreeable because they are different sciences and there is no need to reduce them to one. He adds rightly that both should be eminently reasoned. This is a call to go beyond data and methods of science but that demands proper formation in reasoned discourse. Before scientific quantities and measurement, reason naturally seeks to link them to daily life and beyond. This is what every receiver must dare do, putting reason into use. Jaki quotes William Herschel to emphasize the importance of the use of good reason in doing good science: "Half a dozen experiments made with judgement by a person who reasons well, are worth

---

284 S. Trasancos, *Science was Born of Christianity: The Teachings of Fr. Stanley L. Jaki,* The Habitation of Chimham Publishing Company, Titusville (FL) 2014, 193.

a thousand random observations of insignificant matters of fact."[285] This is a requirement which is either purposely ignored or skipped by modern science. I say this because of the wide spread evidences of constant deterioration in critical thinking in which the human society is engulfed. Important here is the fact that scientific investigations should have strong rational motivations which consequently promote good judgement.

Science needs rigour and exactness not only in experimentations but in reasoning its purpose and consequences. When scientific facts are presented to the unscientific person the greatest problem of our time is that no critical questions are asked, this makes the warfare scientist make sales of books blended on faith and science, without any consistent content except plying on the emotion and ignorance of their readers. Whatever the desire of science, it cannot adequately deal with spiritual issues simply, because they cannot be quantified as Jaki argues, but simply said, they have no tools for it. Science goes a long way to satisfy the curiosity of man about his environment and the universe but it fails to answer the questions that affect human being most, that is about their purpose and end in life and about what follows after matter is destroyed. Man cannot but must reflect on these realities because his mind draws him to them and they matter to him more than the knowledge of acceleration due to gravity or the speed at which the earth turns around its orbit or the knowledge of light years. As the human person reasons his scientific findings, he is naturally driven by the power of his mind to see beyond mere things something greater. That is his power of transcendence. Jaki asserts in relation to culture that:

> Real culture must include an intense attention to the questions that most agitate human beings [...] at the bottom of all psychological anxiety there lies a failure to resolve questions that are distinctly religious in character. Religion cannot, however, exist without taking the form of a cult. Healthy human nature must therefore include a cultic religious attitude as well.[286]

It is that rational capacity that drives the human person to the practice of cult in his religion and this is the reason why we have not yet seen other

---

285 S. L. Jaki, *The Road of Science and the Ways to God,* Real View Books, Port Huron (MI) 2005, 110.

286 Id., *Mind's Matter: An intellectual Biography,* William B. Eerdmans Publishing Company, Grand Rapids (MI) 2002, 13.

species forming associations or making tools or recording their languages into books, or worshipping some supreme being, this is because they have everything the human person has except that self-consciousness that makes man transcends his physical nature and the physical world. The practice of cult is not therefore archaic or something for those who are not intelligent. Man is driven naturally and rationally to the awe of that which lies beyond himself. That greatness which physical science cannot fathom and yet so real that the human mind cannot ignore without falling into self-delusion. If man does not remain faithful to the traditional cult and religions, his negation of it immediately creates a cult that he believes in. This is because the mind naturally seeks some higher and stable truths.

Jaki never got tired of emphasizing the importance of good philosophy, if faith and science are to keep their parallel questions for the good of the investigator who is a reasonable being. He traces various philosophers and philosophical waves that parted from true love of wisdom. Idealism with Hegel, for example, drove man out of the true world and made him loose touch with the base, which is concrete reality that remains true foundation of science and of the spirit, if we are to speak in Hegelian terms. For Jaki Hegelian idealism failed both science and faith because they could not get outside themselves to concrete things outside their thoughts. This is why Jaki continuously asserts that science cannot be done a priori because it must depart from real things and not from thoughts. The Hegelians departed from their own minds and returned to their own minds without reaching God and without doing exact science that demands touch with the concrete things. The Hegelian dialectics destroyed both the base and the destiny they intended to arrive in pure spirit. Their God was their thoughts and their minds. Jaki contents that "Copernicus, Galileo and Newton might have been unfortunate in having been born before Hegel, but the historical fact is that making their discoveries they saw unfolding lawfulness of nature not the traces of their own minds but the vestiges of the creator's mind."[287] This is the difference between a thinker who has God outside his mind and the one who reduces Him to his mind as a product. The true God worshipped by the Christians is not produced by the brain cells of the

287 Id., *The Road of Science and the Ways to God,* 142.

human being or any operation within it but that same God is the producer of the brain of the rational being. Making God a product of the brain is making a product greater than its producer. The failure to arrive to the true God has always been in history the failure of science as well and robust philosophy to trace a good path of metaphysics and transcendence.

According to Jaki, Nietzsche did not make it any better for both science and faith and his philosophy too abandons the honest love of wisdom. Nietzsche's philosophy according to Jaki was anti-science, because he denies the origin of creation on which operates science and therefore placed the cart before the horse: "Faith in the Biblical notion of creation was for him the most fatal aberration of the human mind, and he identified the moment of scientific truth with the rejection of the doctrine of creation."[288] Nietzsche believed in the eternal recurrence and this made any scientific discovery nothing near but always what has been and will be. Science in this sense becomes a futile search not worth taking. Nietzsche in believing in eternal return threw out both science and God from rigorous investigations and relapsed into nihilism.

The rise of positivism was another extreme that reduced everything to the power of a microscope. A rigorous use of reason would point to realities that lie beyond science. Jaki conjures that: "While the objectivity of nature and the ability of the mind to grasp it ever more comprehensively are assertions that conflict with rigorous empiricism, their fruitfulness for the creative scientific mind suggests a natural alliance between science and natural theology."[289] To Jaki, scientific discoveries are a clear manifestation of the rationality of science, because the scientist investigates more and more into nature prompted by reason to take a particular consideration. He affirms that "once that closer look is achieved, it is easy to see that it is not respect for science but dislike of metaphysics that prevents logical positivists and the historicist historian of science from considering discoveries as acts of understanding, showing reason at its best."[290] Reason at its best does not stop on empirical evidences but seeks to understand what they lead to. An historicist is one who sustains historicism, in the most simple

288 *Ibidem.*, p. 143.
289 *Ibidem.*, p. 248.
290 *Ibidem.*

term, a doctrine that history is the basic determinant of human existence and events.

Jaki therefore insists that a good philosophy takes a modest realist position taking real and ordinary things of life's experience seriously. It is from there that science is born and that in awe reaches out to metaphysical realities. The rationality of the universe intimately gives rise to the rationality of science. It is logical that if the universe were irrational, science would not actually be possible and the decoding of revelation would be impossible. Jaki asserts: "faith in the enduring rationality of the cosmos made sense only as long as the world, its laws and (not least) its constants, were given in the deepest ontological sense."[291] The scientist too needs some form of faith [not the same as the Christian faith], whose substance is a belief in himself and his tools, in order to carry out his investigation, as much as the theologian needs his faith, whose content is a belief in God, to understand revealed truth. The bond between the two is their endowment with rationality and the practice of good philosophy to arrive to their desired goals.

When we talk about the rationality of science, we derive it from the understanding that the rationally ordered world cannot produce irrational science. The rationality of science is a rational consequence of the world authored by a rational personal creator God who endowed nature with an infinitely rational law. A good philosophy that bridges science and faith must not be reduced to empty logic. Science and faith are actually like two pillars of a bridge of human effort to understand nature and purpose in the universe with reason as the unifying factor and the intellect as the building mortar. Trying to mix up the two into one is actually like seeking to build a bridge on one pillar. Philosophy seems to have a central role to play in every discipline; otherwise, the search for truth remains a sophistic attempt to win without a race. Jaki puts it very convincingly, "both science and religion are governed by philosophy."[292] Science is a search into nature to discover the truth about it and this cannot be possible without the use of reason and having a reason to do it. Discoveries are looking for more behind what is seen and this should always be driven by the intention

---

291 *Ibidem.*, p. 180.

292 Id., *Mind's Matter: An Intellectual Autobiography*, 209.

to make life better. It is not reasonable or even natural to seek for what is dangerous to human life or a means of making human life more miserable as motivation for technological discovery. The natural human inclination is to make life better and happier and this should not be an exception for scientific investigations.

3.3 Rationality and Natural Theology

Jaki seems to criticize the Aristotelian world view from a contemporary viewpoint without taking into consideration that Aristotle was a child of his time. When arguing about the concept of the universe held by Aristotle, he points out that some obvious conceptions should have been rejected right away by putting into use critical reason. He also argues that Aristotle himself should have realised that it was unreasonable to argue thus. He contends with the arguments Aristotle provides for the difference between wind and air. Jaki argues that accepting such a view marked a moment in which "organismic physics had to side with popular beliefs and reject reasonable views."[293] The point here is not whether Aristotle was right or wrong but that sometimes the neglect to put reason to the rigour it deserves leads to the acceptance of stereotypes. According to Jaki, if major breakthroughs were to be realised in science, there was need to break away from the entire worldview proposed by Aristotle including his theory of motion. Scientific discoveries demand some risks to go against a commonly accepted world view and thus a strong adherence to reason more than stereotypes. There was therefore the need for Newton to break through and provide a tool for the mastery of nature by his lows of motion. This opened the cosmos to a better rational investigation that went beyond the popular belief and mixture of pantheism and science. I would like to assert that Jaki's criticism of Aristotelian world view raises some fundamental problems that many scholars and scientists fall into. The first problem is that it is methodologically wrong to criticise an author outside the context of his time. Secondly, Aristotelian world view could have been wrong, but as all scientific discoveries are, it was the best explanation in the time of Aristotle especially that he gave a revolutionary concept of reality breaking

---

293 Id., *The Relevance of Physics*, Scottish Academic Press, Edinburgh, 1992, 26.

away from the Platonic world view. Perhaps we can learn from the caution of Herbert Butterfield, whose work was not foreign to Jaki that "little progress can be made if we think of the older studies as merely a case of bad science or if we imagine that only the achievements of the scientists in very recent times are worthy of serious attention at the present day."[294] I would like to assert that every epoch and every great thinker is to be judged according to the context of his time and more to that, in trying to refute their world views, humanity is ushered into a completely new world view. Something is always wrong because there is something right to prove it so. There is no guarantee that certain scientific facts we hold as true today may not be obsolete in some years to come.

Looking at the cosmos and how it is ordered for a purpose cannot but provoke some rationality without implying anything beyond. If singularity cannot be denied, then rationality becomes the procedure that cannot be swept under the carpet. Jaki notes that "a historian of science can only note that whatever beginnings the scientific record shows, they invariably indicate a high degree of consciousness." That self-awareness that there exist multiple accounts of realities that affect human history and knowledge is a basic requirement for any investigator. This raises some pertinent questions that we would only want to ask without intending to provide answers for them: What would a blind universe be like? Would reason exclude that consciousness for which self-awareness is possible? Would it be possible for science to operate without putting its data to some coherent structure that make sense? Would unreasonable science produce reasonable results? It is truly arguable that "the origin of a design is an intelligent act animated by purpose."[295] Purpose seems to be a rational product that denying would defeat the process of thought itself and therefore alienating man from his basic operations of consciousness. In order for scientific theories and discoveries to be what they are, they must not be obvious to what their data philosophically imply. This demands that they respect what their limits are.

The issue of limit has appeared several times in our discussion, because it is ultimately what rationality demands in all disciplines and at all levels.

294 H. Butterfield, *The Origins of Modern Science,* Free Press, New York, 1957, 9–10.

295 S. L. Jaki, *The Road of Science and the Ways to God,* 280.

When a particular discipline meets another one in the course of their investigations, the implication is that either they are at an intersection point from which they should define their different paths or that one has passed its limits and should stop and allow the competent other to proceed from there. It may also mean that they all have reached their limits and there is need to develop something new to attend to the new challenge at the intersection. It cannot be reasonable to conclude that it is the end of knowledge, but rather that it is the end of a known possibility at that point, basing on the tools and methods at hand. Knowing one's limits gives an opportunity to see reality with different perspectives that enriches knowledge and its acquisition. There are philosophical connections to reality that demands a philosophical mind and a rational explanation that most times a scientist is incapable of making. We end up most times in conflict because the scientists try to incompetently meddle in philosophical discourses and when the going gets tough, the quickest and safest refuge is to appeal to subjectivism which blocks any further reasonable debate.

Every reasoned discourse collapses when reason is bracketed for some purpose and in this case to avoid facing metaphysical questions or even to avoid recognising that there are other methods that take over from where my own ends. To Jaki "the only kinds of chaos which are helpful to science are forms embodying singular specifics expressive of a design which in form points to purpose."[296] To deny purpose is actually a negative way of affirming it through purposive chaos. "The elimination of a specific way to God based on design and purpose actually strengthens rather than weakens one's rational hold on any purpose, cosmic or particular. [...] It is that rational choice of a rationally coherent set of boundary conditions that make them appear as design."[297] Natural theology is therefore naturally born from the rationality of nature itself, that drives the mind to investigate what lies beyond the world of senses. The fact that nature seems to rationally move towards an end alludes to the fact that it is purpose driven.

296 *Ibidem.*, p. 287.
297 *Ibidem.*, p. 292.

3.4 Faith, Free Will and Purpose

In treating the rationality of science and Christian faith, we need to examine free will and purpose which are ingredients of both but with different results. Faith demands for freedom of the will that determines good and bad acts which are in turn one's reward and punishment respectively. The reward or punishment determines man's actions now in view of the ultimate destiny implied by the immortality of the soul and its salvation. Science raises serious questions about the possibility of free will of the human person with what is termed as physicalism. Physicalism maintains that we are completely explainable in physical terms, that we are nothing more than complicated biological machines. If human beings are just machines, then they just follow what they are designed by their own nature to do. Here already physicalists beg the question about who the designer could be. It is possible that human operations and actions can be explained mechanically but this does not necessarily mean that human actions and operations are mechanical. Good philosophy will dictate at all times that possibility and actuality are two different things. Therefore, the mechanical explanation of human actions does not make it at the same time mechanical.

The basic contention behind this is to remove every moral responsibility for human actions advocated for by religion. If we are pre-determined in the mode of a programmed machine, then our actions are amoral because we have no option to act otherwise, except in the way in which we are predisposed. Proponents of this view, from astronomical point of view, hold that the cosmos is just a big clock-like machine. Rene Descartes and Blaise Pascal could not help but assert that human beings operate like machines with the exception that human beings have a spiritual soul. It is worth noting that Descartes and Paschal were careful to use the term "like machines" and not machines but this comparison will not last in history and other thinkers will soon assert that human beings are machines with flesh and blood. In the enlightenment period extremists like Thomas Hobbes, La Mettrie and others went further to deny the existence of the soul and freedom altogether.

The understanding that human beings are mere machines, gained strength in the world of computers, that some thinkers maintain that machines will be able to act exactly like the human person following the laws of logic. This is true when a machine is logically programmed to follow a sequence of commands that computers are capable of. The gap however remains, that computers would never behave the way they do unless they are commanded or moved by another, or programmed to do so. At the end of every computerised activity there is a free being who has the duty to freely push the right batons. One can quickly acknowledge the limits of this robotic approach by observing a self-parking car that is programmed to see obstacles in the parking space and read the parking lines but does not recognize, for example, a pit that can make the car to overturn. This is because it is assumed that parking spaces have no pits and any change in it, like a land slide in the night, can only be recognised by a free being whose knowledge emerges in the condition in which he finds himself. If a water pump is programmed to run for one hour, it will always run for one hour even if the tank is full and water is overflowing or a loan sprinkler programmed to run at a specific time in the evening, will automatically run even when it is raining until it is reprogrammed to behave otherwise by an intelligent and rational being. It is therefore, an ill-conceived idea to think that a robot can exactly replace a human being. The advancement of artificial intelligence is however, a revolution that cannot be stopped and demands a thorough study to evaluate its consequences. Jaki's *Brain, Mind and Computers* (1969), treats some of these issues underscoring the fact that man cannot be reduced to or compared to a robot. There is an ontological dimension of the human person that put him beyond mere machines.[298] Even where machines replace human labour, they remain machines deprived of the power of volition and ultimately judgement.

Physicalism raises serious questions about the existence of the soul and its immortality and also the substitution of human labour by robots that are computerised to do what the human person would do and perhaps even doing it more efficiently. But then, this is a mixture of biology, technology and physical sciences. The machine made of flesh which is the human person arguably has the capacity to refuse to work or work slowly or stop work before time or even ask questions to a third party because he has the freedom of choice. As machines do not relate to their products, the human person seems to relate very much to what he produces and produces

---

298 Cf. Id., *Brain, Mind and Computers*, Herder and Herder, New York 1969.

them for a purpose that mostly serves him and his kind. The reductionists mode, reducing everything to material reality, would not explain away human freedom and purpose because as much as machines are made in the mode of human operations, human beings do not resemble machines but machines resemble human beings because they are a product of the human mind. We cannot compare human beings to machine but we can compare machines to human beings. Doing the opposite is trying to make a product greater than the producer. This is also valid for the relationship between the creator and his creatures.

Newtonian physics did not make things any better as it demonstrated motion in the universe to an extent that it made results predetermined leading to the philosophical wave of determinism. Determinism means that not only purely physical events but conscious human choices too are the inevitable consequence of their antecedent conditions.[299] It ultimately denies the possibility of freedom of action even to rational beings. It can be true that antecedent conditions may push me to author a book but I remain free to write a good or a bad book. I may be conditioned to write but the quality of my writing is my choice. With Newton, science was able to efficiently predict results of motion of the planets and bodies in it with incredible precisions. The laws that governed electricity and magnetism all demonstrated how mechanically the cosmos operated. "The universality and determinism of the laws of physics gave rise to the belief that the entire universe is a complete and closed system of cause and effects."[300] This implied that the human person and his brain are governed by the laws of physics. We are stimulated to act in a certain way when we are affected by the physical laws and thus we are not free, proponents would argue. Jaki asserts that the success of Newtonian physics "were taken for a proof that mental and volitional acts follow one another in the same ironclad sequence which, according to that physics, regulate the material world."[301] We already note an illegitimate extension of the understanding of the material world to

299 Cf. Id., *Patterns or Principles and Other Essays*, the Intercollegiate Studies Institute, Bryn Mawr (PA) 114.

300 S. M. Barr, *The Believing Scientist*, William B. Eerdmans Publishing Company, Grand Rapids (MI) 2016, 79.

301 S. L. Jaki, *Patterns or Principles and Other Essays*, 114.

explain the immaterial operations of the conscious being. This, as we have already noted before, is a methodological problem that many scientists fall into in their zeal to promote science as the explanation of all things. One would wonder whether those scientists would be acting freely when they argue that human actions are determined and governed by the laws of physics. Newton himself and Maxwell did not agree to the causal closure but the idea gained momentum after them. Subsequent thinkers, notable among them Francis Crick and Edward O. Wilson, began to assert that freedom is an illusion.[302] This however raises countless question about the human moral act. If we are not free, then we are not responsible and therefore our actions have no purpose and are indifferent. This would in the same wake put into crises many facets of religion and ultimately faith. But we see that human beings are either guilty or gratified by their actions and that they act for both remote and immediate purposes. Can the laws of physics, quantities and motion predetermine the actions of the human person? Does a person of faith do good because he is predetermined like a programmed machine or because he is conscious about his good acts and believes it is good to do good and bad to do evil? Is his promised reward for doing good and punishment for doing bad an illusion? Science itself has attempted to answers these questions even though qualitative aspects of things lie outside their competence.

The development of quantum physics by the 1920s seems to have saved physical determinism from the denial of freedom because it called for probabilistic predictions of scientific phenomena. This demonstrated that there is no point that scientific tools can gather enough data to exactly predict, in the mode of determinism a subsequent phenomenon. We cannot at any point say that we have enough data to predict exactly what will happen in the future basing on statistical information at hand. This accrues from the nature of science itself, being non definitive, but the best explanation of physical phenomena at a given time, science can have no claim for a definitive and exact prediction of the future in a mode of making it predetermined. There must always be a probability that the contrary may occur however small that variation may be. Science is always and at all times associated

---

302 Cf. S. M. Barr, *The Believing Scientist*, 79.

in its discoveries and theories by a certain percentage or degrees of error. Even quantum mechanics cannot explain away the pitfall of physicalism because what goes on in the human brain that determines his actions cannot be governed by physical laws. It was until 1931 that Gödel theorem dealt a death blow to the mechanistic and formalistic capacity of scientific theories. The Gödel theorem implied that "the power of reason, even in the bloodless realm of mechanics, involves something more than machine-like-behaviour. It involves understanding of meaning and judgements of truth and falsehood."[303] The above faculties cannot be governed mechanically by physical laws even when they are sometimes provoked by physical sensations. Science itself has a proof for this fact. According to Giambattista Formica, in the most general sense, Gödel theory confirms two important things, on one hand that mathematics would contain even in its simplest parts an undecidable truth and therefore indemonstrable, on the other hand, that coherent science cannot prove its coherence.[304] Scientific theories are therefore plausible explanations and not strictly proofs of their certainty. Human behaviour cannot be reduced to a purely mechanical operation because the human person is a free being. One should however cautiously apply Gödel's theorems within its context and logic to avoid misinterpretation or application out of context. Formica notes that Gödel was wrongly interpreted when he was still alive and there are several wrong interpretations that come out of the lack of proper comprehension of his incompleteness theorems.[305] In the final analysis, human actions cannot be governed, determined or judged using mathematical or mechanistic laws.

As science engages in the investigation into what constitutes nature, scientists are also free to act within freedom and responsibility to achieve their end, which is the production of theories about nature. The man of faith has truth as his destiny while the scientist has facts, its accumulation and generalisations into theories as his final end. They both employ intellect and reason to achieve their objectives. For a Christian, his freedom makes

303 *Ibidem.*, p. 85.

304 Cf. G. Formica, "L'assiomatizzazione della matematica e le ricadute dei teoremi di Gödel", in P. Pecere (a cura di), *Il libro della natura*, vol. 2, Carocci, Roma 2015, 249.

305 Cf. *Ibidem.*, pp. 251–254.

him a person and responsible for all his actions that gives him a sense of purpose and sense of his religion that rewards him for being good and punishes him for being bad. As science looks at all these is a mechanistic and deterministic way, religion looks at the same reality in a purposeful and spiritual way. Purpose has occupied philosophers right from Thales to our modern times that some philosophers have chosen to deny it basing on some findings of science. Socrates refusal to escape from prison indicated a strong conviction for a purpose that is not mechanistic in nature and advocated for the immortality of the soul even when his understanding of the soul was not what we have now. Socrates made his conviction in the form of an argument that:

> Since everything moved for a purpose, that is had a soul, therefore man, acting for a purpose also had to have a soul commensurate with that ultimate form of purpose, which is to gain personal immortality further, he also intimated that once one entered immortality, one either enjoyed the ultimate benefits of a good conscience or suffered forever the consequences of having died with a guilty conscience.[306]

This argument of Socrates, from the mouth of Plato, is a purely rational and before the Christian time, that conscience must dread bad acts.[307] One does not need to be a Christian to know that a good act gratifies the actor and a bad act brings guilt and shame. We can argue that science too must have a good end guided by good reason. Our modern times have seen the abuse of science like the use of atomic bombs, biological and chemical weapons and even the suspicions that surrounds the current pandemic of the coronavirus, codified COVID-19.[308] As in the Christian faith one who does good things is rewarded with eternal life and one who does bad by hell.

---

306 S. L. JAKI, *Late Awakening and other Essays,* Real View Books, New Hope (KY) 2006, 119.

307 Cf. PLATO, *Phaedo*, 69e–84b. (translated by ROUSE W. H. Denham, edited by WARMINGTON H. Eric – ROUSE G. Philip, New American Library, New York, 1956).

308 There are undeniable contentions that the "Corona Virus Disease of 2019" (COVID-19) could have been produced in, and intentionally leaked out of the laboratory in Wuhan. These are polemics that I would not like to enter into, but serves as an indicator that science can be used for irrational and selfish ends.

So should scientists dread death with guilty consciences for not directing science towards the good. Jaki argues that:

> Socrates was certainly a most unusual person. He distinguished himself in hand-to-hand battle, the only battle known to man before he introduced proxy fighting, first by canon and musket and lately by cruise missiles. To make farce out of tragedy, man armed by remote control weapons, boasts of personal heroism.[309]

The advancement of technology because of science has not spared the military field to the extent that powerful countries no longer need airbases for air strikes on countries thousands of miles away. The use of drones in wars provoke many ethical issues that brings to question the purpose of science. A purposeless science will definitely militate against man himself. However, Socrates purpose according to Jaki did not promote science, he thus argues:

> Unfortunately, it has not been remembered in philosophical circles, or at least not taken in its full significance, that in standing up for purpose, Socrates also dealt a mortal blow to a healthy perception of how we know purpose. Very unintentionally, but also most effectively, he became a counter-reductionists, by fighting reductionism. [...] While the Ionians *physikoi* and their successors, the atomists and Anaxagoras, reduced everything to matter and motion, Socrates reduced everything to purpose.[310]

Consequently, the physical universe, taken as an organism lost its position as an object of scientific inquiry and teleology became an empty motion towards an end without a purpose. This is why, according to Jaki, from Descartes purpose became a mockery until a breakthrough by Newton's physics.[311] According to Jaki, Darwin chattered purpose by his natural selection that theoretically occurred without purpose.

When we talk of purpose we want to refer to final course, which is acting for a purpose. Purpose is therefore closely related to ontology, if there is no ontological causality, we cannot dare talk about theological causality. Jaki held firmly that:

> Only within such realism [acknowledging what is and its cause] can both freedom and complete ontological determinism be upheld with consistency, because both

309 S. L. Jaki, *Late Awakening and Other Essays*, 119.
310 *Ibidem.*, pp. 119–120. Cf. Id., *Pattern or Principles and Other Essays*, 50–62.
311 Cf. Id., *Late Awakening and Other Essays*, 120.

> will be seen as created by the One who alone in his ontological infinity can call forth ontologically finite entities into existence.[312]

To avoid all polemics of subjectivism, we bear in mind that man's sense of purpose is very much rooted in his being a subject to some higher power. Man is at all times subject to his creator God. Jaki argues that:

> But that subjectivist pedestal for purpose is obtained after purpose had been dethroned from its position as something ontologically objective. That ontological objectivity does not lessen a whit just because man knows of purpose through immediate self-reflection on what he is doing when he acts, for he cannot act consciously except for a purpose. This is something most objectively evident, although it is also most evident subjectively as well.[313]

As man is conscious that he is an acting being, he will always act for a purpose dictated by his consciousness. The purpose for which he acts may be qualified as good or bad, right or wrong. It is what distinguishes him from all other beings and when in the name of science some people deny faith, the unexpressed purpose is to destroy religion and affirming purpose as such. Since science deals with quantities, it provides a step towards ontology (not phenomenology), ontology remains the foundation of purpose and a good philosophy. Jaki asserts that:

> Quantities are in the forefront when we begin to know, which is to know something, a thing. Recognizing of objects must therefore be the first step for a philosophy that takes pride in objectivity. Objects, their registering, must come first, so that among other things we may speak objectively of some other first steps in philosophy, such as free will, causality and purpose.[314]

Jaki criticizes phenomenology as an avoidance of ontology and rationalism as a run away from objective reality and advocates for a realist point of departure from concrete objects, if we are to have any meaningful Christian philosophy or theology. He cautions against reductionism and subjectivism that has become the fashion of our modern times. "Now if purpose is a primary irreducibility, one must avoid any discourse about it that imply some quantitatively verifiable aspects. This is so because the truth of that aspect would not then depend on its philosophical support, but on measuring of

---

312 Id., *Pattern or Principles and Other Essays*, 130.
313 Id., *Late Awakening and Other Essays*, 124.
314 *Ibidem.*, p. 130.

that quantitative aspect."[315] Quantities are therefore only stepping stones towards a concrete understanding of purpose in the universe. We cannot therefore escape purpose if we are to be reasonable in trying to understand the universe and its content. "Purpose is a redux, though not in the reductionist sense, but in the sense that it re-introduces itself whether one likes it or not."[316] Purpose is therefore related to free will when it is purged of its mechanistic nature and applied to free acting being.

Man is a purpose driven animal and every time he achieves one goal he sets another one, either because he has gained new insights or that he encountered problems he did not expect. But there is a purpose that lies beyond physical existence and human set goals and this is what every human mind struggles to grasp to the point of asking the purpose of his own existence.

> More than the merely physical is implied in the fact that through every action made for a purpose man wants to become something more. The most appealing purposeful actions have always reflected man's urge to rise above himself as if his true weight gravitated upwards to a higher realm of imponderable intuition.[317]

Such is yet another distinguishing mark of the human person as opposed to a machine. He believes in realities that are beyond himself and his reason and intellect assure him of truth beyond himself and that such a truth remains reasonable and not a fantasy. Jaki argues that neither Kant nor Darwin could destroy intellectually purpose in their own exposition which could make their own works purposeless not to talk about transcendental meaning and content of purpose. To Jaki, the human intellect "cannot accept the destruction of reason by reasoning."[318] This is why human beings cannot be confined to the universe in which he finds himself and reason to contradict itself is for it to fall into irrationality per excellence.

Our motive for discussing purpose in relation to the rationality of science and faith, is to point out that the two cannot be reduced into one and yet they all serve to give a holistic vision of reality. When science claims

---

315 *Ibidem.*, p. 131.
316 *Ibidem.*, p. 132.
317 Id., *The Saviour of Science*, William B. Eerdmans Company, Grand Rapids (MI) 2000, 131.
318 *Ibidem.*, p. 140.

to demonstrate that there is no purpose in the universe, and therefor there is no benevolent creator, it tries to reduce the irreducible to science that is concerned with quantities and their measurements. In the same way Christian faith cannot be reduced or interpreted according to the discoveries of science because the purpose of science is not the purpose of religion. Denial of purpose in whatever form whatsoever, denies the very act of its denial as purposeless. For every conscious action must have a purpose. Jaki argues that "it should be easy to see that such a conviction is a truth [...]. For whenever such a conviction is assailed by doubts that exist not merely in the doubter's mind but are conveyed to others, the assault on purpose by being purposeful, testifies to purpose itself."[319] Man is separate from all other beings because he is conscious, meaning that he knows his actions and therefore knows why he acts in a particular way. Purpose and consciousness cannot therefore be separated. Extended, to the universe one cannot fail to recognize that the details found in it fit in and out for a purpose that the mind will never at one-point exhaust. Our cosmos would not be there without exceedingly specific quantities being present from the genesis of the cosmic processes. It is what science recognizes now as anthropic principle which goes far beyond the mechanistic operation in the universe to a purposeful one.

> From its earliest stages the universe headed towards a state which alone can by its physical properties account for two developments: One is the formation of stars around which alone planetary systems can exist; and the other is cosmic preponderance of carbon in the proportion necessary for the emergence and maintenance of life.[320]

The precondition does not reflect a self-existing or auto generated cosmos but alludes to the cause of that principle that the cosmos seems to follow.

Denying purpose tantamount to denying freedom and causality which encourages rational sceptics and irrational deniers to reject in the name of science both religion and science itself. Despite all the arguments of science and the evolution theory promoted by Darwin and his followers there still remain astonishing details of the natural order that provokes some great

---

319 Id., *Means to Message,* William B. Eerdmans Publishing Company, Grand Rapids (MI) 1999, 87.

320 Id., *The Saviour of Science,* 133.

awe to the human mind. It can be asserted that without the freedom of man, religion is not possible, because it stems from his higher desire to relate to metaphysical realities and only rational being is capable of having a religion and doing science. Darwinist will always find two challenges with Catholic thinkers; the creation of the universe in time by a personal creator and creation of individual souls out of nothing. Reason helps the rational being to make choices and determine right and wrong without being mechanically driven. Science surely has sensation as its point of departure but science cannot explain or account for spiritual realities like desire, empathy, shame, guilt, joy, anger and the existence of the soul; realities that are not governed by physical laws. The denial of it would lead science back to the pitfall of reductionism. Scientists have also no ground to blame the people who have faith from holding these spiritual realities as real without proofs. This is plausible because not every knowledge should in any case be proven in the scientific way, aware that science cannot even prove the truths of the contents of their beliefs outside quantity and numbers. Besides, such a proof would require a different method other than that which empirical sciences are accustomed to.

It can be argued that if the universe rationally demonstrates some purpose that alludes to a rational and free creator, and if man is free and not radically determined to act the way he acts, then the scientists ought to have some ethical considerations when they engage in scientific investigations. These ethical considerations do not stem from religious precepts but from reason itself. This will occupy us in the next section.

3.5 Ethics and Science

It takes rationality and an honest use of it to show that science is limited in conclusions to some degree of error and will continue to do so because of the nature of what it is. The truth will always elude experiments and have explanations and foundation in metaphysics. Quantum theory, in the 1920s, has already undermined physical determinism by demonstrating that scientific findings are only probable as we have seen above. It is therefore difficult to determine scientifically a scientific law. Stephen Barr explains with the example of the decay of the nucleus thus: "The laws of physics do not allow one to predict exactly when an unstable particle or nucleus will decay; they only allow one to assign a probability for it to happen within a certain period of time. This probability is always stated in terms of the 'half-life'."[321] This can be extended to all physical processes. Barr affirms that, "however, it should be kept in mind that all [physical] processes are, strictly speaking, governed by the principles of quantum theory."[322] To understand that science cannot provide a base for judgement we need to examine ethical values that are dictated by man's ability to have reasons for his actions. This is why we cannot also attribute man's freedom in the same way as the unpredictability of quantum theory. We do not know how man may act in all situations but we know that if he is to act freely and responsibly, some rational processes are required for his actions. This is technically called deliberation, that involve his reason, intellect and power of volition. Actions, be it scientific or religious, are only called human when they bear some ethical consequences that point to and defines the rationality behind every human endeavour. That humans are free is undeniable as long as the human mind operates using reason, intellect and will and as long as we make choices from divergent alternatives. Jaki puts it thus: "There has not been and never shall be a better and more inescapable proof of free will than one's immediate experience of it, which is applicable even to determinists arguing against it."[323] As soon as one is confronted with a choice, one immediately experiences the reality of personal freedom. We acknowledge that one's choices from all diversity of possibilities depend very much on how much one has developed his power of reason, his culture and his environment. But we also assert that despite all these conditions, man can still choose what he wants to do freely, because his culture dictates onto him the obligation to think deeply about his destiny and to make a choice.

The rationality of nature and therefore science cannot be separated from the human consciousness about ethical values imbedded in the very nature of humanity. "Truth and morality are intimately connected."[324] Jaki argues that Francis Bacon already remarked most accurately that "the practice of

321 S. M. Barr, *Modern Physics and Ancient Faith,* University of Notre Dame Press, Notre Dame (IN) 2003, 177.

322 *Ibidem.*

323 S. L. Jaki, *Pattern or Principles and Other Essays*, 137.

324 Id., *The Road of Science and the Ways to God,* 305.

science would be governed by sound reason and true religion."[325] There is something important to be considered here that seems to be lost in the present day, that is, the use of sound reason which is ultimately having right motives pursuant to all that we do. Sound reason is actually equal to honest reason that acknowledges the limits of our abilities. This, when added to true religion, brings the human being back to the same light of reason that lies beyond itself. True religion can only be defined by the revealed religion in which God is a person, revealed, rational and relational. It is not a fabrication of the human mind, unlike the laws of the cosmos, that has to be discerned and formulated by human science. Were it to be a human fabrication, indeed, such a mind would itself qualify to be supernatural!

Revealed religion cannot therefore be relegated simply as superstition in which man surrenders to irrational forces beyond his reach. True religion is faith having a strong rational justification. Jaki argues thus: "Clearly, science has ethical dimension and if the road of science is leading to the ways of God, this has to be also true of the connection of science with ethics."[326] This means that the scientist cannot be indifferent to the scientific experiments he performs if he is to remain honest to himself. For Jaki, it should be "an honesty which is not a monetary compliance with disconnected data but a free and lasting surrender to their consistency obviously puts on it a philosophical quality which removes to a higher level the apparent conflict between freedom of the will and law of nature."[327] Ethics should put the scientists on the path in which it is not pecuniary value that matters but the ethical considerations of scientific investigations. There are scientific practices that cannot be promoted because they directly contradict the purpose of good science, for example, the use of stem cells to cure cancer patients or experimentations using human beings as specimens or production of vaccines using particles of murdered babies. This raises ethical questions because it involves saving one's life by destroying another or using another person as a means to an end. Questions regarding scientific

325 *Ibidem.*, p. 306.
326 *Ibidem.*, p. 307.
327 *Ibidem.*, p. 309.

technicality and ethics can be extensively discussed by bioethics.[328] I would like to point out here simply that every scientist is bound in conscience to act rightly and according to good reason and to seek the promotion of the common good.

For science to take root in this world, it had to battle with the irrationality of an undeveloped world to the end that modern science cannot fall back to the irrationality that it fought. This calls for, in Jaki's words, "resolute break with subjectivist, irrational world views and the acceptance of the consistency of nature and of the consistent exercise of man's freedom. Science will march freely only as long as it marches responsibly."[329] This seems to be the core of the argument for any rationality in science. Science cannot be formed into a game of a blind and indifferent or amoral end as long as there is a human being behind every scientific investigation. If there is responsibility in science, then there are choices a scientist must make and making such choices calls for an ethical foundation that can be arrived at by reason. It is worth bearing in mind that due to high specialisation, scientist may not realise the ethical implications of their engagements. Modern man is most of the time engaged in production of parts, whose whole [complete product] is only known by the top management who may unethically use them. One may not know or may undervalue the ethical implications of his involvement in the production of a bomb, or may not even know at all that he is involved in a production of a destructive weapon, when his involvement is simply cutting iron fillings which form the most dangerous part of the bomb. He may not comprehend that the fatality of a bomb is actually based on the quality of its bombshells made of those iron fillings. Every scientist is ultimately responsible for his product and its use both directly and indirectly.

We cannot take nature for granted because it has an author who is rational. According to Jaki, "both that freedom and that responsibility will keep pointing and in a coherent manner beyond science to a realm where its road turns into the way to God."[330] We can only arrive to a conclusion

---

328 Cf. Id., *The Gist of Catholicism and other Essays,* Real View Books, Pinckney (MI) 2001, 159–182.

329 Id., *The Road of Science and the Ways to God*, 309.

330 *Ibidem.*, p. 312.

as this when we commit ourselves to the demands of a rigorous thinking which is rationality itself. The use of reason does not exclude the use of logic as long as we depart from an honest truth and continue desiring for it. Science has developed over the centuries because of adherence to some facets that were true but also because of the openness to new discoveries. This openness is an act of freedom and only a rational being can perform such an act. Science therefore becomes a rational endeavour because first and foremost the world is rational and secondly, the human being uses his reason to interrogate it.

3.6 Science and Faith

Jaki demonstrates how science enters into play when God and faith is demanded. "But science can bear witness about the supreme platform, the universe, from which alone a rational inference can be made to God, the Creator."[331] The cosmological argument remains one of the gigantic evidences of God's operation in creation, supported most of the time by the orders that prevail in the universe. The more science digs into the universe, the more it becomes alive through the depth of its design. Scientific cosmology cannot but bows down to such evidences that speak for themselves. Besides, if science is possible, it is because the cosmos is available for its operation and the cosmos has a law that the scientist cannot account for – the legislative that Christian religion calls God.[332]

Jaki argues that modern science bears witness to the universe in two ways: "firstly, science has re-established the intellectual dignity of the notion of the universe and, secondly, science has disclosed a stunning degree of specificity about the universe."[333] These have made the discourse about the universe more transparent, real and realistic. This continues to provoke wonder in the scientists and the beneficiaries of science and when taken seriously keeps reason from the extreme of the rationalists. There is something concrete about the universe that we are dealing with and yet there is something that continuously lies beyond what we ordinarily see. So much

---

331 Id., *The Saviour of Science*, 223.
332 *Ibidem.*
333 *Ibidem.*

of the universe was already known by revelation for which scientific discoveries are just a verification with some specific details. This also alludes, in the thinking of Jaki, to the contingency of the universe; elements of the universe reveal that it is not a necessary existence. This conclusion would not then be far-fetched "if, however, such is the case, then it can be shown that the universe is that supreme reality that cannot have within itself the reason for its own existence."[334] What then accounts for its existence? This is the point from which science enters into dialogue with faith because it (faith) accounts for the necessary existence – God who is the author of this order and all that exist in the universe. From scientific specificity arises the question why the universe should exist the way it exists and not in any other mode. It should evoke awe in the researcher as to why the universe should demonstrate understandable order and not any other disorder. It is even more challenging to scientists who attribute everything to randomness and chaos: why should there be order out of randomness and chaos for which science is possible? Even more challenging, why should there be at all, if there is, randomness and chaos and not nothing? These are questions stemming from a reasonable science but that goes beyond the tools of science into metaphysics and the philosopher's field. Theology cannot come in at this point until mediated by the philosophers for a full comprehension of the revealed truth. This makes revelation to have its foot on reality knowable by physical means and in an analogical way by reason. It is only arguing in a logical and reasonable way that we can arrive at the understanding that science and faith lie within separate conceptual domains that do not contradict each other. In Jaki's words, "so much for the argument that alone can convict secular man in the court of reason, that his claim about God's non-existence is a misuse of reason, whatever else it may be."[335] It can be argued that a fallen reason is redeemable while a corrupted one is doomed. This is because when reason is corrupted, then the source of the liberation of the human person from folly is jeopardised and consequently imprisoned. Perhaps it is worth quoting the analysis of Ashley that "secular humanism tends to make too sharp distinction between the philosophies based on

334 *Ibidem.*
335 *Ibidem.*, p. 224.

reason, and theologies based on faith in revelation, which it rejects."[336] These rightly find themselves into a host of problems because every philosophy and religion develop within a culture from which it cannot be alienated without contradictions.

I would like to assert that the laxity in putting reason to a rigorous use, not in the rationalist way, will always create conflict where they should not be and bring confrontations when none is called for. This according to Jaki is caused by philosophical shallowness which actually is the concern of our modern time that is preoccupied with how things function and not why things function the way they do.[337] It seems that the ordinary person of our time is not ready to settle for anything stable, because flashes of information and images are dinned and congested in his mind so much that rejecting and filtering becomes a cumbersome business to be let loose to the battle of specialists whose testimonies are subjectively embraced. If it is science, it is taken without questions assuming that every scientist measure to his name and if it is religious, it is dismissed as dogmatic and bigotry, that can be embraced by those who want to or rather those who are irrational. Such is the philosophical shallowness of our time that puts into bracket reason to be comfortable.

Jaki recommends that in the age of science the survival and thriving strategy is the employment of good philosophy or metaphysics and a thorough respect for reasoning in depth. Reasoning in depth should imply the desire to embrace the truth, an inclination which leads to faith as the ascent of the intellect towards the absolute truth and thus a metaphysical foundation on which all truth stand. If God is not looked upon above all as a creator of all who is ready to be tried in the court of reason, he will be degraded into a miracle monger.[338] An investigation into the scientific understanding of miracles according to Jaki would form a good area of a scientific research for which I will not venture into right now. Worthy of mention, because of its philosophical content is the argument of Jaki that God does perform miracles that are understandable to the human mind because miracles are principally meant to enhance faith. In fact, Jaki in a somewhat sarcastic

336 B. M. Ashley, *The Way Towards Wisdom*, EPUB e-book.
337 S. L. Jaki, *The Saviour of Science*, p. 225.
338 *Ibidem.*

way asserts that "miracles are not for the benefit of scientists as such or of Academies of science. If they were, they would be tricks of the devil, who in turn would not fail to make appear plain fools all those who had fallen in his trap."[339] It therefore follows that miracles are for those who have the faith to comprehend it and that they do not necessarily go against the law of nature but make the law of nature operate in a particular situation in a way that it would not naturally occur at the same time and for a specific purpose. God, being the author of natural law does not contradict himself and thus "miracles are supernatural interventions in nature."[340] Jaki points out how Darwin made a radical change in his faith as a result of shallow or no philosophy at all. He asserts: "Charles Darwin, shortly after he returned from his voyage with the Beagle, he turned suddenly from a Bible-wearing fundamentalists Christian into a dedicated enemy of Christianity as well as of creation."[341] This is probably because he did not understand with profundity the Biblical context including the various episodes of miracles and creation helped by sound science and good philosophy.

I would like to argue that if reason is taken with the seriousness it deserves and without being selective in its use, religion, based on faith and reason, will always help the human person ascend to metaphysical realms without contradicting with modern science, whose methods permits only quantitative investigation into natural things in motion.

3.7 Incompleteness of Science, Precision and Faith

Physical reality is so vast that no single reductive explanation or study would do it justice. The only justice we can do to the universe as created by God is to accept that we can do as much within the camp of our competence and remember that there always remains a possibility of doing more at all times depending on our level of intellectual evolution and tools. Notable however is that every camp contributes to a better understanding of what this complete represents. Various branches of ancient and modern physics/

---

339 Id., *Bible and Science,* Christendom Press, Front Royal (VA) 1999, 201.
340 *Ibidem.,* p. 203.
341 Id., *The Saviour of Science,* 226.

science point to something ever new all the time and never definitive at the same time.

> For just as the notions of organism and mechanism failed to prove themselves as the final word in the conceptualization of physics, no different is the case with its mathematical formation. Successful and wide-ranging as this may be, no particular formulation of it can claim to itself on purely intrinsic grounds the glory of being the true representation of the structure of physical reality.[342]

In a nutshell, one can only claim complete knowledge and means to knowing physical reality that science deals with if one is the author of it. This would apparently be trying to do the impossible since one cannot bring anything into existence out of nothing. For Jaki, "metaphysical questions are therefore not out of place because it provides another way of understanding the universe and physical reality."[343] It is arguable that it is philosophy that unites all the ocean of knowledge of the universe begging for its author. This however remains a constant research into nature with no anticipation or expectation that one day we shall have a definitive knowledge of existing reality under the docent of empirical science. Such a knowledge only true Christian faith can guarantee because, apart from natural reason, it depends on revelation as well. On the continuous search Jaki argues:

> Furthermore, as there is no telling what unsuspected experimental data will turn up in the course of physical research, there can be not strict scientific basis to claim that the "greatest of all claims", as Einstein called the physical system of the greatest conceivable unity will one day be achieved by science.[344]

We contend that as human knowledge matures, more light will always be casts on what has previously been known that will affect both the understanding of science and faith. There should therefore be an openness to scientific progress and flexibility without forgetting the unchanging ontological foundation on which it is based and progresses.

When we consider from such as "strict scientific basis", Jaki wants to point out that many claims are made in the name of science, and that those claims are not scientific and are slippery grounds for such claims. "No one would dare assume today that there is nothing new for man to observe in

---

342 *Ibidem.,* p. 121.
343 *Ibidem.*
344 Id., *The Relevance of Physics*, 124.

the physical world."[345] The complexity of the physical world makes a scientist appear as a blind man with layers of scales in his eyes and each time a scale is peeled off he sees new reality never seen before and sees what was seen before in a relatively different way. Nature is an everlasting challenge that has to be constantly unmasked. The process of unmasking requires various methods and the acceptance that one single method for all things is not enough. The philosophical method allows the view of nature from various angles to converge to that one point of an ontological reality that provokes thinking and reasoning.

Mathematical physics was so successful in developing Pythagoras vision of the world of numbers that many scientists believed the universe would be explained by geometry and numbers until all was set ablaze by Kurt Gödel incompleteness. He proved that formalism of the *Principia Mathematica* or any other formed system that is not too narrow cannot have in itself its proof of consistency.[346] This remains a fundamental challenge to every scientific theory because it lays down theories and has no capacity to prove its own validity within itself. Theories are therefore theories and will always remain incomplete and they require another theory to prove their consistency. Gödel showed that the propositions of a theory cannot be proven or disproven within formalism.[347] There is therefore no guarantee of the internal consistency of a scientific theory. Jaki argues that Gödel's proof can only be a great help for those who honour limits. In his own words he says: "For the speculative physicist this implies that there are limits to the precision of certainty, that even in the pure thinking of theoretical physics there is a boundary present, as in all other fields of speculations."[348] Jaki adds that this is only good for those in whom rigid rationalism has not killed off the seeds of intellectual humility otherwise despair would be the only escape route for such.[349] It is arguable that science by its very nature is fluid and does not admit of any rigid propositions. Gödel's incompleteness should be both an impetus to continue working but also a warning that

345 *Ibidem.*, p. 131.
346 *Ibidem.*, p. 127.
347 *Ibidem.*, p. 128.
348 *Ibidem.*, p. 129.
349 *Ibidem.*

there are limits to what can possibly be done and therefore, the claim of having once and for all a universal theory of all things would be superfluous.

After many mathematical scientists thought that mathematics was enough to explain physical things in nature, it still turned out that it's limits are insuperable.

> Mathematics can translate only one aspect of things. It can count the atoms in an apple, it can tell whether this number is odd or even, but mathematics cannot provide either the apple or its atoms. Much less can it say anything about what are known as the pleasing features of an apple.[350]

There are elements of physical reality that escape mathematical and scientific tools to the extent that to ask a scientific question about them would tantamount to asking an irrelevant or bluntly a silly question. That it is nice to eat an apple before or after dinner cannot be determined by science especially what "nice" entails, because it is a quality that cannot be expressed quantitatively or even transferred to the feeling of a third party who eats an apple at the same time. The ground therefore remains for other disciplines since science cannot at any time provide a definitive intelligibility of nature.[351] Jaki alludes that "it will not be otherwise with the metaphysical concept of physics, which may very well prove itself the ultimate form of physics as a science, but certainly not the ultimate intelligibility of the physical world."[352] The conclusion of Jaki with these words invites scientists and all other stakeholders who contemplate an understanding of nature to take on appropriate and honest steps in their search: "the great arm of physical science, the overall synthesis of the scientific understanding of the universe will remain for all practical purpose what it has always been, the ever-remote objective of an intellectual faith."[353]

The scientific endeavour explicitly remains within the framework of some form of trust that lurks behind every effort that reality contains more than

350 *Ibidem.*, p. 136.
351 Cf. *Ibidem.*
352 *Ibidem.*, p. 137.
353 *Ibidem.* It would be interesting to have a look at the various concepts of faith. For synthesis about the notion of faith: cf. J. U. Ozioko, *Rationality of the Christian Faith in Richard Swinburne*, Urbaniana University Press, Rome, 2019, 117–124.

what we see and that we are capable of always trying to intellectually interrogate physical reality for a better understanding. This is definitely fundamental for science which may not be the conception of the modern scientist, but it remains a fact that may be given any other name or face. Even if faith is a theological gift in the Christian sense, it forms part of human intellectual activity without which it may be grossly impeded. A scientist must have trust both in his thoughts, theories and tools of investigations notwithstanding the fact that he has to believe that the acquisition of new knowledge is possible. When in history scientists dismissed faith as empty and therefore not true, they did it with a great belief that what they are saying is not just a fact, but truth affirming the possibility of a stable truth. If truth is one and stable, then the negation of faith in a revealed God is the ridicule of the gift of intelligence and reason. This is true not only because it goes against the intellect, but also that it negates an historical fact. The Christian faith is not just an intellectual speculation but a truth based in history that a man called Jesus was born in time, performed wonders that are recorded in books, some of these wonders were qualified as miracles that go beyond science. The fact is that the story of this man Jesus, has changed time and centuries. If believing that he is God, revealed in human form is a matter of faith, accepting that His influence on the world has been great and continues to be great, is an historical fact. It is arguable that the magnitude of his influence, if he is not to be taken as God, can only be attributed to a super-human and therefore, accepting such a glaring fact would be a matter of prudence and honesty. As science demands courage and determination to defend its hypothesis and theory, so does faith demand ascent to divine truth experienced and lived. Jaki affirms that "scientific work needs daring statements, bold conjectures and at times a rather firm devotion to a particular hypothesis. Yet, even more, it demands a clear awareness of the true character of the statements made."[354] It is arguable that devotion to some particular hypothesis may not necessarily be so because they are proven but because of the trust in the intellectual capacity to prove the truth of its content. Some amount of trust both of the scientist's ability and in his tools is indispensable if science is to avoid falling into error. It logically follows

354 *Ibidem.*, p. 148.

also that science does not necessarily operate on verified facts all the time as propagators of scientism would wish to portray. The recognition of this demands full deployment of reason and intellectual rigor.[355]

The mathematization of science is appealed to so much so that it builds an ideology of precision and exactness for the ordinary non-scientist. It is not by chance that the ordinary person does not question scientific assumptions even when they fail to explain a phenomenon. Normally in the ordinary mind the scientist is right at all times and, if wrong, it is because something went wrong. Jaki explains this mentality when he states:

> The use of numbers, formulas and equations always conveys some magical force, particularly for the layman to whom these pages of the famous ninth edition of the encyclopaedia Britannica were written. Such parlance makes him imagine that behind the exactness represented by the prolific use of numbers there is a solidly established experimental evidence and a scientific logic unencumbered by preferences and wishful thinking.[356]

The simple conclusion is that "the scientist knows about these things." More often than not there is no truth or established facts beyond the assumptions dressed in numbers. Formulas most times prove a present conclusion known by the one who generates it. The search for the reality either occupied scientists so much so that, abandoning it, was like an abomination in the religious language. There were neither proofs nor experiments that confirmed any reality of it. This is sometimes the same accusation meted on religion, the lack of proof or evidence. This, however, cannot be used to justify religious beliefs and faith but points to a deeper understanding that reason arrives to certain truths not by mere experimentations that most of the time prove facts arrived to without them.[357] Jaki explains by giving the example of the composition of electrons and how positrons were arrived at first without experimentation: "In the composition of atoms as: electron, protons and neutrons, the missing positive charge, positrons were arrived at by speculations before experiments."[358] The advantage of sciences is that it backs up its speculations by experiments which is a requirement

355 *Ibidem.*
356 *Ibidem.*, p. 166.
357 Cf. *Ibidem.*, pp. 167–171.
358 Cf. *Ibidem.*, pp. 174–175.

of its method. Experiments, however, are designed to produce a preconceived result that is speculated by the scientist in the form of hypothesis. Some form of speculations without initial proof is therefore envisaged for science to grow. This cannot be applied to faith because the proofs of faith experiences would not be by experimentation.

Human reason and intellect therefore live outside any experimental realm. Intuitions cannot be predicted or caused by some manipulation of physical matter such as brain cells or any other scientific conditions. Physical sciences continually change, and a wide range of assumptions are a constituent part of it. We already note that scientific generalizations are made over a wide range of data from a small amount of evidence. Jaki points to the fact that the wide range of data that accounts for minute detail crystalizes to a generalisation whose derivation is always forgotten; he asserts thus, "yet one can never be sufficiently aware of the facts that orders of nature expressed in the various formulas of physics is but the redirection of given sets of data and formulas to more general ones."[359] He warns that the hope for a science that definitively defines matter is an illusion that is very much unscientific.[360] Science should be contented at all times with the parts that they are capable of investigating and not with the totality that they cannot say anything about.

Discussing the cosmos and the struggle of the scientist to understand the Milky Way, Jaki contends that "science is not about the unknown, but about what is known. It has to aim at the ordering of available data, not at speculations about the unknown."[361] In his lifetime, he struggled to move science away from any a priori processes because it would make science engage into what it should not. Besides, a priori thinking cannot form science according to Jaki, because it reduces it to man's thinking and brain and therefore loses the rationality that governs the cosmos. The consequence of using a priori argument is that it neither constitutes science nor reason. For Jaki, it produces irrationality and fantasy. Science should be an investigation into nature and not an imposition of the mind on to nature. To attempt to scientifically explain origin into something outside creation

359 *Ibidem.*, p. 186.
360 Cf. *Ibidem.*
361 *Ibidem.*, p. 209.

would imply imposing our minds on nature and not taking it for what it is. The daunting fact that history cannot deny is that science has not, with all its advancement, succeeded in establishing that living beings are a result of some inorganic chemical reaction. Creation seems to be reasonably unsurpassable even in the world of science. Jaki argues thus:

> Clearly mathematics was meant as a tribunal beyond which no court of appeals could be imagined. As a result, a rather bitter pill had to be swallowed by those who saw organic and intelligent life as the inevitable product of chemical evolution and who therefore would have felt far more comfortable with the belief in living beings popping up spontaneously in every nook and cranny of the universe.[362]

Existence seems to owe its explanation outside material things, to the creator who willed that they begin to exist by the act that Christian religion calls creation.

The history of science demonstrates various levels of focus and development: from an organism to mechanism to mathematics to atomism and to the search of precision that affects science in all ways. According to Jaki:

> Aiming at ever greater precision in measurement affects physics in several ways: it may establish and confirm a theory, it may pose the need of further clarification, it may reveal unforeseen cracks, and at times it may force the abandonment of a theory and impose a search for a new one.[363]

This makes science or physics a very unstable body of knowledge that cannot claim authority for a long time. As observations improve with tools, so does science become more accurate and yet not definitively accurate. Science therefore has nothing to prove that faith holds as true and believable. As science progresses by proof within the limits of the tools available and degree of error, faith is a stable acknowledgement that some metaphysical reality lies beyond nature that has no need for a proof or experiments. The ascent to faith is never by proofs but by deep religious experience that cannot actually be fully expressed or transferred to a third party. Jaki points to how we naturally get in touch with what is outside us writing, "we are very certain that we do elicit sensory activities which in a real but

362 *Ibidem.*, p. 229.
363 *Ibidem.*, p. 247.

mysterious way put us in touch with objects in the world outside us."[364] This is a natural window to metaphysics that follows naturally from our knowledge of physics as long as man has the exclusive capacity to think and reasonably organize his thoughts into a reasonable body called knowledge. To conclude that the feeling of God cannot be proven and therefore not be real is the same as denying the assertion of someone who claims happiness because we cannot experiment and determine the degree of his happiness. We can only accept another person's joyful disposition by a collection of external data indicative of it. As we cannot transfer the joyful experience to a third party, so that someone else can feel that joy without becoming another, so can we not fully transfer the religious experience that is sustained by faith from one individual to another. The Christian faith we are talking about here is not a product of a natural ascent from nature alone, but a product of revelation in time with a specific language and culture. Science therefore contributes to some understanding of natural elements of nature that has nothing to do with the degree of faith in God and creation as the work of God.

When a Christian holds a firm belief that God created the world and all the things visible and invisible, he does not make a statement of science but of faith in God as the efficient cause of all that is and all that is possible including science itself. Science and all scientific activities are therefore possibilities within the creation of God, because there is nothing that science introduces in being or rather causes to be. It is however all the time evident that "fundamental problems of physics will always raise metaphysical questions."[365] This is simply because man has the capacity to transcend what is in the physical realm to that which is reasonable. Jaki argues that, "actually, not even the positivists' disclaimers of all metaphysics could so far dispense with postulates that one genuinely goes "beyond" physics. Thus, even the most radical positivist method in physics will have to suppose the universe to be such as to fit the method and vice versa."[366] The physicist cannot therefore be confined with the physical reality because the

364 R. J. Connell, *Substance and Modern Science,* The Center for Thomistic Studies, Houston (TX) 1988, 148.
365 S. L. Jaki, *The Relevance of Physics*, 343.
366 *Ibidem.*, p. 343.

knowledge of things themselves is no longer physical. Physical realities and knowledge about them have philosophical implications that the scientist has to leave to the philosopher. Jaki drives the point home by saying: "humble protestations about one's insufficiency in philosophical matters does not remedy one's incompetence."[367] It is however an invitation to acknowledge limitations and not deny the existence of knowledge beyond the physical. Trust and not proof lies at the foundation of scientific investigations and denial of it is like denying science itself.[368] Jaki gives the example of Copernicus as not driven purely by science in his search: "It was a firm trust in something that clearly was not physical but metaphysical that conjured up for Copernicus an entirely new vision of the universe."[369] This alludes to the fact that the faith of the Christian scientists was an impetus for them to have a clear vision of the cosmos and investigate into it. Scientific data in themselves are nothing until they are put into a coherent form not simply in a scientific manner, but in a logical and philosophical manner. Jaki argues:

> The pure rationalist, the extreme positivist, the anti-philosophical physicist, are clearly at a loss when coming face to face with the task of accounting for the constant use of assumptions, the nature of which they try to ignore. Yet, without recourse to philosophy, one is indeed hard put to find justification for the fact without which physicists simply cannot do their work: the unavoidable necessity of translating mathematical formalism into a conceptual system based on sense perception.[370]

Much as the assumptions of science are sometimes overlooked and theories taken for granted by consumers of science, we recall that a degree of error admitted when using a one-meter telescope used from the earth, is not the same as that of the modern time launched in space. Assumptions of science affect results over a long period of time as tools improve. The philosopher is well placed to render scientific data meaningful, because philosophy deals with fundamental principles governing a phenomenon and not specific details of individual sciences. Only philosophers can conceptualize scientific findings within a wide range of knowledge. This gives a holistic view of

367 *Ibidem.*, p. 344.

368 Cf. *Ibidem.*, p. 345.

369 *Ibidem.*, p. 346. Cf. Id., *The Saviour of Science,* 96.

370 *Ibidem.*, p. 353.

reality: physical and metaphysical. A scientist and the theologian are saved from myopia by the philosopher. Jaki argues thus: "clearly man cannot think, speculate, reason or build specious theories even within physics itself, let alone outside it, except in a way in which metaphysical aspects of physical things and phenomena are inextricably present."[371] Human reason has the capacity to transcend the physical and the proof of transcendental reality is actually this capacity of the human mind to move from physical to non-physical, from observables to non-observables in nature. The dialogue between the physical and the metaphysical is therefore inevitable and our simple task is to frame a secure means by which this dialogue can be fruitful for the good of the human person.

3.8 Misconceptions about God and Faith

The hostility that revolves around faith are basically and often founded on misconceptions which end up being taken as truths. Some few examples can illustrate this argument; the first misconception is about God and his nature. Many of those who hold science as a means to demystify God are those who have created their own gods of the mind and placed them before their science to make comparison. They miss the true Christian God and when they levy their attacks on Christians because of their God, they attribute to God what is not of God. Jaki would put it this way: "Whatever man's ambitiousness, anyone who thinks that he or she can do so well that even God cannot do anything better has a very inadequate notion of God."[372] Noticeably, Jaki does not even speak of the knowledge of God to such a person but just a notion of who God is, just to show how grave it is to speak from arrogance or ignorance. It follows that if one does not have the notion of God as an absolute and omniscience being, then one can say anything about that god which is a product of his own mind. It follows necessarily that all arguments based on a mistaken notion of the true God will be false except for that 'god' of the mind of its author.

---

371 *Ibidem.*, p. 358.

372 S. L. Jaki, *The Drama of Quantities,* Real Book Views, Port Huron (MI) 2005, 32.

The second misconception is the understanding of faith, that it is contrary to reason and that people of faith believe at the expense of the good use of reason. Proponents of this actually do not understand that what they call faith is the superstition of the pagans condemned by Christians themselves. The Christian faith at no point admits mediocrity as far as the use of reason is concerned. This is why the Catholic Church has always taught that faith and reason cannot be separated and faith is not the property of the ignorant. Indeed, simple minds cannot arrive at faith which is another form of knowledge. Every Christian arrives at faith because it is a gift from God, but also because it is reasonably sustainable.

The third misconception is about the understanding of mystery. Christian mystery is neither a gap in the natural realm nor an escape from the limits of our human knowledge, but a reality that lies at a higher horizon of human capacity for which our reason is rightly attracted and understood in its nature as mystery.

It can be argued, then, that most arguments that put science and faith in opposition are flawed, based on wrong premises and popular misconceptions. It is unfortunate that even faithful Christians, when challenged by these misconceptions, cannot rise to defend themselves and those who attempt to rise to defend themselves end up defending a "wrong" God or rather a misrepresented true God. In order for one to deny God or any other reality, one must have acquired all understanding and knowledge about such a reality, its existence and or evidences of its non-existence. Otherwise, it would be safer to assert existence basing on the little that can be known, other than non-existence whose defence demands a thorough knowledge of what can be known. This requirement, critically examined, would necessitate that man should be eternal here on earth to be able to acquire all the knowledge about the cosmos visible and non-visible. The reality is that this is not possible because the length of our lives is what it is and the physical limits are endless. Revealed truths are aids to bridge this gap in our capacity; not in the sense of filling in, but because the truths of its content are plausible. The cosmos is globally understood when science and faith give their valuable contributions without anyone claiming supremacy over the other.

The historical fact of the life of Jesus Christ are facts that are evident even from the mundane point of view, because Jesus existed in time and space and his coming to the world has had enormous influence that cannot be ignored. Jaki puts it very vividly when he writes that "if facts are measured by their impacts, the greatest fact of history is Christ by any count. Even on a purely empirical basis, it should be clear that no one made a more massive and more lasting impact on history than He did."[373] It is another issue whether he was/is the Christ. Because of the revelation in Jesus, we have the linear concept of time that was already revealed in the New Testament and that time will be fulfilled in him who took humanity away from the cyclic eternal return. As science studies motion, it acknowledges the fact that no motion is possible without the first unmoved agent. Science has repeatedly tried to establish the beginning of the universe outside creation, and at all-time arrived to the reality that the universe could not merely come out of nothing without a cause. Aristotle, without any revelation had already, in the principle of hylomorphism stated that nothing comes into existence from nothing disappears to nothingness without a trace. Even when this alluded to the eternity of the universe and matter, which the Christian doctrine does not support, it competently pointed to the first cause of the universe and the eternity of the human soul. Aristotle considered the soul as the substantial form of the body.[374]

Both science and religion in their corrupted forms ply on ideologies. A deteriorated form of science is scientism which sees nothing valuable outside science and a corrupted form of religion is fanaticism which reduces everything to irrational faith. All these survive and thrive on ideologies accepted without question and common sense.

Looking at our modern time it is doubtful to validly talk about common sense, because critical thinking, that has its base in common sense, is hidden in the tomb of laziness and indifference. To give credence to common sense, there is need to remove the veil of relativism and indifferentism that have become the hiding places of ignorance. We need to uncover unfounded ideologies with a rigorous use of reason.

373 Id., *Mind's Matter: An Intellectual Autobiography*, 187.

374 Cf. R. J. Connell, *Nature's Causes*, Peter Lang Publishing Inc., New York, 1995. Connell discusses the Aristotelian four causes in the light of Christian understanding of God as the efficient cause, as the cause of the other three causes and more so not as an agent but the author of cause itself.

3.9 Faith, Reason and Science

The Catholic Church has held in high esteem the use of reason in understanding faith and promoting a friendly relationship between science and religion. The church has time and again offered guidance to scientists and not opposed or impeded the growth of science as argued out several times by Jaki. The various papal addresses to the Pontifical Academy of Science are testimonies to this fact. The encyclical *Fides et Ratio* of John Paul II is a monumental guide to the pursuit of a deep understanding of faith motivated by reason and can serve as a guide for the understanding of faith and science. We would like to discuss how the Church has always guided the path of science in history.

Jaki was named as an honorary member in the Pontifical Academy of Science by Pope John Paul II in September, 1992 and contributed a lot to a balanced view of science and faith among the academia.[375] It is partly because of his contributions that I would like to discuss critically faith, reason and science. If we are to gauge Jaki's contributions to Catholic Christian thoughts in the light of faith and reason, we could classify him as an objective scholar who knows how to give credit where it is due and discredit where it should be. Jaki was never afraid in his exposition to criticise even the highest ecclesiastical authorities with a non-destructive mind pointing out how better things could have been handled to avoid clashes. His strong conviction from evidences that science could not be born except in the Christian Western Civilisation and with a firm root in the medieval time is an incontestable truth. This truth has historical evidences that demands a monumental work to dismantle it. It is not out of place that this provokes negative reactions from both Christian and non-Christian scholars who want to keep the status quo.[376] Haffner advices that:

> Jaki's historical approach should greatly help Christians to overcome an inferiority complex in considering the relation between faith and science. His work could also help them avoid the pitfall of developing a false sense of superiority which is bred by the cultivation of facile concordism between those fields.[377]

---

375 Cf. P. Haffner, *Creation and Scientific Creativity: A Study in the Thoughts of S. L. Jaki,* 17–18.

376 Cf. *Ibidem.,* pp. 215–223, for specific contributions of Jaki to the life of the intellectual life of the Catholic Church.

377 *Ibidem.,* p. 222.

Whoever reads Jaki's work should always be prepared to discern well his arguments, which when well understood is a scientific, philosophical and theological tool in the hands. The polemics that come with his arguments are basically inevitable, because of the nature of the material that he treated and the passion with which he treated them.

I would like to point out from the beginning of this section that the Catholic Church has in history sought to promote science not without misunderstanding explicitly with scientists like Galileo Galilei. Even in our modern times, the Catholic Church still struggles with the moral aspects of scientific advancements and the advancement in technology due to science. We note that the moral position and the philosophy of the Catholic Church will always provoke more hostilities in a secularised world interested in the battle to free the world from all sorts of religion and religious beliefs.

Jaki argues that a proper understanding of religion and faith should be cultivated for all Christians, so that they can clearly defend themselves on matters of faith and science and ultimately God. He asserts thus: "As long as one means the ultimate ground of existence when using the word God, religion will lie behind any effort to fathom the meaning of human existence."[378] The thorny argument that calls for a strong foundation of faith guided by reason asked by science has always been the origin of the universe and man's place in it. It follows that, once the mainstream religions are neglected as explanations for the origin of the universe and man's meaning and purpose in life, the propagators of science would substitute it with another form of religion whose creed fits their mind set. Religion is born naturally from the desire of man to have meaning to his existence and reconnect to his God. In the Christian perspective, apart from the dictate of reason, revelation in time and history gives him the strongest foundation of his faith in a God who created everything and continues to sustain his creation. Science also looks at existence but in a different way using the same reason. To proceed to decipher the content of faith or science, human rationality is inevitable.

---

378 S. L. Jaki, *Questions on Science and Religion*, Real View Books, Pinckney (MI) 2004, 129.

Jaki continues to argue that Christians are most of the time challenged by those who advocate that we may not be alone in the world; there could be some extra-terrestrial beings like us or even better than us. He argues that the search for similar beings as human on other planets raises questions about faith that should not be ignored. These questions according to Jaki deserve no answers until there are proofs that there exists extra-terrestrial intelligence. Christians have no obligation to defend themselves against speculations. In any case the all-powerful God could create as many things as he wanted without being less God. Jaki continues to argue that "this means that delights as one may in some preconceived ideas about existence, one has to allow the facts of existence to prevail over one's ideas about it."[379] This therefore requires that faith be guided by reason that does not allow scientific questions to be mixed with those of faith, on a clear rational ground that they lie within separate conceptual domains.

He continues to argue that honest science and theology must admit priority of facts, if it is to have a secure ground for its arguments. The philosopher is important at this point because he begins in his epistemology from how man can know and what he knows depends on the fact that he has to know something and not nothing. This something is not a speculative something, it is real existence. Jaki asserts: "To know without knowing something is at best a dream of German idealism, but most likely a sign of psychological deterioration."[380] This is valid for both science and theology because the knowledge of God that is a product of the mind like that of Kant is deism, and yet true science according to Jaki was nurtured in the theistic environment in which God is not a product of the mind but a person. "Whether man, modern man likes it or not, he cannot wean himself away from his Christian moorings. Just as facts control all good science they should control all sane reflection in religion as well."[381] That the earth is no longer the centre of the universe, or that man is not, is matter of fact and science. This, however, does not make man less what he is because "the centre is man's mind as riveted on facts, which he can use as a springboard for knowledge which is one, is nonfactual yet wholly

379 *Ibidem.*, p. 164.
380 *Ibidem.*
381 *Ibidem.*, pp. 164–165.

rational."[382] To have faith therefore means also to have the ability to put within perspective the findings of science guided by faith and reason. The place of man in the universe is a struggle of man to understand himself because he is capable of doing such a reflection. Whatever place man puts himself in the universe does not affect the nature of God, who created him including his rational ability.

In *The Gist of Catholicism* (2001), Jaki affirms that the encyclical *Fides at Ratio* of John Paul II rightly puts Christ at its centre and as the centre of the Christian life. One would wonder what this has to do with science from the Christian perspective. Jaki argues that there is a ray that runs from Christ to science and back to Him.[383] Jaki's arguments seem to be directed to Kant's assertion of Jesus as being merely a good teacher of morality. In contrast, Jaki affirms that reason can only admit the divinity of Jesus, because he said it and the events surrounding him testifies to it. Jaki argues that the revelation of God as "I am" and the affirmation of Jesus that "before Abraham was I am" is the greatest of the divine word ever uttered.[384] From his moderate realist position, Jaki argues that such affirmation could not have been made in form of an idea or hypothesis or a wishful thinking. It points to the unrelenting question of existence and authorship of the universe beyond the physical material. Jaki emphatically asserts: "clearly if science owes its very birth to Christian faith, indeed to faith in Christ, science cannot really be foreign to the matrix that gave it birth in the first place."[385] The encyclical *Fides et Ratio* points to the fact that the separation of faith and reason only makes it run the risk of making faith superstitious and reason empty. Faith and reason according to the encyclical should be closely linked and this serves the same purpose for science and faith. Science without reason will fall into scientism with all the reductionism related to it.[386] The Pope points out to the danger of scientism as a departure from the proper use of reason that has occupied man throughout history: "This

---

382 *Ibidem.*

383 Id., *The Gist of Catholicism,* Real View Books, Pinckney (MI) 2001, 39.

384 *Ibidem.,* p. 41.

385 *Ibidem.,* p. 49.

386 Cf. John Paul II, Encyclical Letter *Fides et Ratio* (14 September 1998), nn. 48, 88.

leads to the impoverishment of human thought, which no longer addresses the ultimate problems which the human being, as the *animal rationale*, has pondered constantly from the beginning of time."[387] The Pope points out to the ethical requirements for scientific endeavours that go beyond technicality. There are many technological achievements of science, but not all that is technically possible is ethical and admissible. It is through the employment of reason that one can discern between what is both technically and ethically correct. Ethical correctness cannot be provided by science and yet it is indispensable for it to remain a rational endeavour.

Jaki argues that the relation of faith, reason and science is ultimately a moral issue. This is because it calls to wisdom which science does not provide.[388] This is important in the argument of Jaki because one cannot arrive at wisdom, which is a moral requirement, without acquiring facts about science and his faith. This affirms our constant argument in this chapter for the need for a rigorous use of reason. There must be a rationale behind every human endeavour. It is therefore recommended by Jaki that the advancement of science be learnt and ethically used by Christians discerning well the negative and positive effects. Such, demands for some basic formation in science, so that scientific claims are not taken or rejected at their face values. We have already seen in the proceeding section that the scientists cannot be ethically indifferent. Jaki argues from a theological point of view that ethical questions will have to be considered at all times, because of the effect of the original sin on the intellect and the will. Jaki characterises modern time thus:

> When that information historical and speculative has been acquired, one can, but only then, begin talking of the wisdom which science cannot deliver but for which science clamours more loudly than ever, in this age which becomes more and more the age of scientific giants an ethical dwarf.[389]

It can be asserted that for an equilibrium to be reached between scientific advancement and ethical considerations, focus should be placed on an objective use of reason that promotes a good philosophy. The world is truly becoming everyday a scientific giant and an ethical dwarf. This is

387 *Ibidem.*, n. 88.
388 Id., *The Gist of Catholicism*, 50.
389 *Ibidem.*, p. 51.

because scientist would want to apply the quantitative measurement to justify ethical actions or the lack of it. When it comes to right and wrong, scientific data and statistics cannot pass the verdict because something is not right merely because the majority hold it to be so but should rather be because it is objectively justifiable as right. It would be more secure to look for sufficient reason and not sufficient sentiments as a justification of our actions. It is lamentable however that our modern time tends towards what is emotionally acceptable and not what is reasonably just. These challenges will occupy us in the next section.

3.10 The Challenges of Our Time

Jaki was greatly concerned about the challenges that have descended over the modern generation who can be described as tepid in all spheres, spiritual and intellectual. The 21st century has so many challenges that range from the adoration of science without even knowing what science actually is and what it concerns, the multiplication of religions that pretend to value nature without acknowledging the author of nature itself, outright weak atheism that is more of a fashion and populism than serious objections to real existence of a true God and to the uncountable ideologies that live in the virtual social world. Such is the new world order that faces a philosopher with countless questions. In treating rationality as the distinguishing trait of the human person, we cannot but have to face the reality of our time. Jaki talks of a God-less religion as a new form of pantheism that has invaded the modern man. The modern man believes in what is popular and he is allowed to follow his feelings and not his reason. It is enough to investigate into modern relativism to demonstrate this fact. "Man and woman of our time who care to be intellectually honest may prompt some readers to admit, at least to themselves that their boasting about not having a religion may be a mere camouflage for their avid espousal of the religion known as secularism."[390] We live in times when the human beings live in open contradictions without remorse. The secularised world has invaded every culture and it is attempting even to create a universal culture with a loss of identity to belong to any particular group of people or even place. It has

390 Id., *Patterns or Principles and Other Essays*, 66.

created a situation in which being objective tantamount to being offensive and provokes the claim of elitism and yet according to Jaki, "the impotency of science to provide norms and goals in addition to tools is dawning on humanity at a time when modern scientific culture is definitely discrediting the faith in the Creator as anachronism from a naïve age."[391] The norms needed for science is definitely the call to a moral orientation towards the goal intended by scientists in their investigations. The impotency to evaluate the necessity of the first step makes science a risky venture if not subjected to the rational rigour required of it. This is a formidable challenge of our time that has to be faced by both theologians and philosophers if our knowledge should be of service to the world in which we live.

For Stephen Barr, our times is characterized by prophets of pointlessness like Richard Dawkins, whose publications he describes thus: "His pulpit is an endowed chair in the 'Public Understanding of Science' at Oxford, and the message he proclaims in his elegantly written, if somewhat waspish, books and articles is that the universe and life have no meaning".[392] In writing about the meaninglessness of the universe, he however intends to convey a meaningful message from the universe itself. Our times are characterized by arguments, sometimes by highly specialized fellows with no philosophical background. They draw the wide majority behind them, because they promise liberation from a slavery that does not exist. Our time needs scientists, in the work of Barr, "who have as well the joy of their faith, which tells them that the beauty of nature points to something higher, to a wisdom greater than their own".[393] It is an act of honesty to acknowledge what lies beyond us and outside us that can neither be reduced to our mind or limited to our thinking. Good use of reason should lead to that. Barr wraps it up that "faith does not need to wait upon the latest research."[394] As a gift from God, faith is based on a stable truth that all that exist, visible and invisible, are products of a free God who freely caused them and sustains them in being.

391 Id., *Science and Creation*, 357.
392 S. Barr, *Believing Scientist,* 183.
393 *Ibidem.,* p. 186.
394 *Ibidem.,* p. 189.

At this point we ought to bear in mind that our definition of the universe is a totality of things that operate in their diversities without contradiction, put more succinctly as Jaki did: "The totality of things in motion." It therefore follows that one would not be talking of the same universe when one talks about it as an organism or as a mechanism for which God's wisdom and intelligence cannot be evoked. Modern thinkers are required to define their terms as they advance their arguments.

Another challenge of our modern time that Jaki demonstrate, in his attempt to reconstruct the history of science, is the attempt to read into history and not allow facts of history to speak for themselves. Jaki, in the course of his research, demonstrated how scholars read their own minds into history and deliberately ignore some glaring contradictions that time and history has laid bare. This was one of the reasons that made him bring to light the works of Pierre Duhem, *Scientist and Catholic* (1991), *Reluctant Heroine: The Life and Works of Helen Duhem* (1992), and several other books and essays. Jaki himself experienced first-hand selective reading when in conferences his learned colleagues demonstrated the ignorance of his works. The most glaring example he gives is on the exposition on Gödel incompleteness of scientific theories.[395] Good science that is rational and that acknowledges the rationality of faith departs from an objective search for the truth. As we read history we should bear in mind the time of the thinker and what motivated his thoughts. We cannot say Aristotle's' world view is a complete non-sense. It may appear so now because of our modern tools, but it was a great deal of facts at his time that set into motion philosophical and scientific investigations.

Barr laments on the folly of our modern time among experts saying:

> The problem in some any other fields is that not a few of the experts give every appearance of being kooks themselves. Architectural experts destroy beautiful old Catholic churches. Liturgical experts give us liturgies that are painful and embarrassing ordeals. Literary theorists write impenetrable prose. Psychologists give us Twinkie defences. The question arises: When is one to trust the experts and when is one to trust one's own instincts? It may be the central problem of our time.[396]

---

395 Cf. S. L. Jaki, *A Late Awakening and other Essays*, 1–12.
396 S. M. Barr, *The Believing Scientist*, 211.

To arrive to ask such a question one must be critical, a characteristic that lacks in our time. The discernment of the fleet of information that floods our minds every day must be the preoccupation of modern thinkers. This will help us to reformulate Barr's question thus "when is one to trust experts and when is one to trust his own reason?" To trust one's own reason, it is paramount that reason be cultivated and put into a rigorous use that it deserves. The dilemma is that many find it onerous to be critical and therefore depend on what google can offer as an answer.

Another challenge of our time is what Jaki points out as the desire to be all inclusive. Our modern time has learned to replace objective truth with politically correct language. This has built in the attitude of not offending anyone even when they are in the wrong. The ideology is that what one feels should not be judged or condemned even when it goes against moral standards generally accepted. It is enough to examine the ideology of gender to drive this point home. That the human person cannot sufficiently be identified as male or female, is a form of irrationality that must be taken seriously. Religious convictions are dismissed as religious bigotry. The irony is that whoever holds his faith in high esteem is publicly condemned, but not allowed to condemn. He is forced to listen to all others but not allowed to express religious views in the name of not imposing his religion on others. Secularism and science can preach their religion, but traditional religion, and especially the Catholic religion, is to remain quiet or be persecuted. Unfortunately, in the face of all these Catholic intellectuals and sometimes hierarchy have chosen to be indifferent or to play it safe. Good reason demands for equity in the exercise of the freedom of expression and religious choices. In real life it is not possible to be all inclusive without losing one's identity. What we need is unity in diversity. Any form of mediocrity undermines catholicity as a system of objective truth.[397]

Jaki then invites us back to good philosophy when discussing non-Thomist Thomism.[398] He asserts thus: "Good philosophy is both natural

---

397 S. L. Jaki, *A late Awakening and other Essays,* 227.

398 By non-Thomist Thomism, Jaki asserts firstly that his philosophy may appear very much non-Thomistic but that his philosophical procedure is Thomistic. Secondly that his philosophy is not Thomism understood in its various branches. Doing a non-Thomist Thomism is therefore focusing on the epistemological dicta of Thomas Aquinas. Cf. *Ibidem.*, pp. 224–227.

and supernatural in its inspiration because it has to remain a love of facts in their whole range. That range must include the facts whereby supernatural revelation becomes known to man."[399] The bottom line of this invitation is the objective search for truth from facts that are present to us in real life. This is the meaning of "facts in their whole range". A good philosopher or theologian as scientist cannot afford to be selective when it comes to facts without contradicting reason and his very mission. We can choose our fields of studies, tools and methods but we cannot accept facts selectively, because we want to accommodate all. We trust experts when they are objective and distrust them when they are not. It is doing this that the love for facts that naturally grow into the love of wisdom, which is philosophy in a nutshell.

Jaki describes our modern society as a sick society. He thus emphatically asserts: "Actually, the condition of modern western society is far worse. It is the condition where the sickness or rather abnormally is taken for a sign of health."[400] He continues to describe our modern society as a parasite driven by secularized society, one in which man is self-sufficient with no need for the sacred or divine. The divine is replaced my material possession and pecuniary value. Standards of life are set by what is comfortable and popular. The rule of life is set by the majority whether they are right or wrong. The major engagement of secularism is the elimination of everything to do with religion and religious believes.[401] Religion should be strictly a private affairs whose practice in public is detestable. According to Jaki, the French and German face of the "new religion is Christianity without Christ. Or whenever Christ is included, He is included only as a man but never as a saviour."[402] The insistence of the elimination of religious signs like crucifixes from public places speak for themselves. In fighting religion, our secularized world forgets that the first target to hit is western culture and civilisation that has been built for centuries. It is a boomerang on the very fabric of the western life. The modern society is busy editing constitutions of the founding fathers of nations to remove any references made to God. This is done to accommodate non-believers in God. Reason would dictate

399 *Ibidem.*, p. 230.
400 *Ibidem.*, p. 149.
401 Cf. *Ibidem.*, p. 150.
402 *Ibidem.*

that the word God would not affect atheists, since it has no meaning for them and therefore no need for the fuse to accommodate them.

Returning to the parasitical analogy of Jaki, the west lives a parasitical life in the sense that western civilisation was built and nurtured by one of the oldest religions called Christianity. According to Jaki the parasite which is secularized governments feed on their host and as it is always, it will at one point die with him as no hookworm survives its host. Perhaps it important to quote this entire text that calls the human society back to the value of the family and religious values. "Families are the living organism of society and especially in a society which has largely turned parasitic. Should most families become parasites, the final phase of a visible disintegration of society would be in sight."[403] This could in a few years be called the prophesy of Jaki, because the trend of things seems to go fast in that direction. It is high time to ponder on the question of Gilson about the West: "Can a social order, begotten by a common faith in the value of a certain principle keep on living when all faith in these principles is lost?"[404] According to Gilson the Western civilisation was built on the principle that held a firm belief in the eminent dignity of man. When the prominence of the dignity of man is lost, the western world stands the danger of losing all its long-accumulated values and prestige and the question, asked many years ago by Gilson whether it will continue living is something to be pondered in every honest heart.

Our modern time therefore does not only shoot its own foot by destroying religion and culture but dismantles the family that should stand for its perpetuation. A society where moral relativism reigns with acute rise in crime and the dismantling of stable families is probably heading to an unfortunate end. This may appear a very grey picture of our modern western society but it is the reality that should be faced with serenity placing science and religion under the guidance of good reason and the common good. There is need to rise up to great standards and strike the necessary balance in modern life.

---

403 *Ibidem.*, p. 152.

404 E. GILSON, *The Unity of Philosophical Experience*, 219.

Our modern society is sick according to Jaki because it is driven by activism. We would have a long list of activists if we were to name them. It should however be born in mind in the words of Jaki that "activists are intellectuals who do not want to see a difference between sane and insane ideas."[405] When something is right, it does not call for activism, because it will defend itself against all odds by being right and when reason is put into proper use, then a critical analysis of every situation would expose its folly or justify its worth. When a matter is subjected to activism, it is clear that either the activists have no rational justification for their advocacy or that the whole system lacks the rationale for proper judgement and therefore must coerced into good practice. In both circumstances, activism would not be treating a problem but its symptoms. Jaki agrees that these challenges make the future quite grey and he presents a possible remedy to it thus, "when a human enterprise is at a crisis point, the most helpful remedy consists in tracing the road to the present, as the diagnosis of the symptoms of a disease is basically dependent on a careful analysis of the past record."[406] By asserting this, Jaki wants to demonstrate the importance of a philosophical analysis of history that exposes the truth which is most times obscured either deliberately by the politics of science, popular ideology or involuntarily by ignorance. Science whose rationality accrues from nature cannot advance to great heights if it abandons the rigorous use of reason that promotes an honest investigation into nature.

3.11 Transcending Science and Materialism

In looking at the rationality of science and Christian faith, we came to the realisation that science has its own limits, even though both use reason to achieve different objectives. Jaki argues that science should be contented with quantities and its measurement expressed in numbers. The two camps are always distinct and only philosophy can hover over to understand them. Science has enormous potentiality that should be seen within its scope and the limits. Jaki argues that scientist will not go in pension soon, because what nature provides is so much that scientists until now have

405 *Ibidem.*, p. 153.
406 Id., *Science and Creation*, 359.

not discovered much of it and even when scientists will come with an overall theory explaining natural reality, they can never be sure it is the final theory because of Gödel's theorems. As long as science operates on what is given and it is the activity of man, it cannot but have to admit the fact that the world and nature cannot be understood without asking metaphysical questions. The never-failing expression of Jaki is that whoever does not want to do metaphysics should not even dare open his mouth to talk. Beginning to talk, for example is already to have a purpose, and to have a purpose is to will to articulate something, and to will to articulate something calls for an adequate language for it that should have some logical sequences. All these then already fall outside quantities and numbers into metaphysical and philosophical concepts. It is true Metaphysical questions are therefore unavoidable as a matter of the dictate of reason itself. It is apparent therefore, that beyond the lenses of the microscope lies a greater reality to be contemplated. The marvel behind and beyond the natural is contemplated because our reason is capable of delving in it.

In the spirit of intellectual honesty and respecting boundaries, scientists in admitting metaphysical questions meet the theologian and above all the philosopher who has the duty, by the nature of his discipline, to put together universal principle that provides better understanding of both science and theology. This however does not mean that theology or philosophy is a continuation of science. It means rather that they have tools and methods that can further the investigations into what nature crudely provides. We have a wealth of knowledge to gain from science, even when we do not have to take science as prove for the content of faith. We are enriched by the mathematical-experimental science, because by analogy we can arrive to the superior science of the spirit moving from what is provided by nature, to what lies beyond the physical, that is the ultimate cause of all things, revealed in the Christian religion and faith as God, the creator.

This is a point in which the two flow like a separate stream with philosophy hovering over them. Because sometimes the human desire supersedes the honest search for knowledge, there is bound to be moments of clashes that should always be amicably resolved. Since the common point of departure of both science and religion is the concrete reality provided by nature, the two could be reconciled in the human intellect and reason. This is because the mind is capable of comprehending simultaneously the two separate domains without having to alienate itself.

The consequence of this is the conceptual springboard from which nature is understood and appreciated with its author. The meeting of science and religion in the same man should be a moment of self-reflection and self-building. The convertibility of scientific and mathematical knowledge to real life of the world is a great evidence of this harmony. Jaki warns that in understanding of faith and science, theologians should avoid a blind reference to science simply because it corresponds to some Biblical records because the two have different formal objects. Time has passed and many things that affect our knowledge have changed. Theologians can benefit from the achievements of science to enlarge their horizon of the deposits of nature. There is need to embrace the achievements of modern science with a clear mental picture of what they serve. A denial of the achievements of science by religion would lead into unpopular holding of archaic positions illumined by science, not to forget the obvious which is holding on to superstitious belief. Science should sharpen our imagination to enter more deeply in the marvels of God's creation. When we talk of imagination, we want to mean a lucid mental picture of reality under the guidance of true and honest reason. The general principle already proposed by St. Augustine should hold so that, each time there seems to be a clash between what the Bible says and what science has discovered, the Bible should be interpreted accordingly in the light of good reason. Jaki writes: "The reasons behind this were twofold: God was reason incarnate and man was made in the image of God. Therefore, God could not teach anything to man that defied basic laws of reason, though God could and did teach him things that reason alone could ascertain."[407] It is not the knowledge of atoms and molecules that will take man to heaven but his ascend to God in his faith motivated by reason. Scientific findings are therefore only dispensable stepping stones that cannot contradict revealed truths.

There is a continuity from the limits of the scientific capacity, and we can say that the limit of science is not the end of knowledge, but a beginning of

407 Id., *Questions on Science and Religion*, 154–155. Cf. Id., *Bible and Science,* Christendom Press, Front Royal (VA) 1996.

another level of knowledge. Michael Heller, like Jaki, would argue: "The existence of the world, its rationality and the moral choice that underlies all science are but imitations of a horizon that seems far away yet it permeates everything."[408] This penetrating reality is enduring, and yet cannot be fully explained simply by the scientific method. It makes sense, creates wonder and yet it is not a fantasy. It is the mystery that nature carries in itself but which goes beyond itself and calls for another method of investigation.

To transcend science is not just a mere choice, but by the dictates of nature that honestly investigating into it, points to a reality that can only be explained metaphysically by philosophers. At this level, the scientist has to acknowledge another method that is not experimental in order to appreciate the philosophical method that deals with universal principles. The philosopher is on the other hand to appreciate the endeavour of the scientist in demystifying nature in order to clear it of apparent mysteries and superstitions, so as to confront the wonder of the true mystery that has as its point of departure ontological questions. Jaki affirms from a realist point of view that "in philosophy one either begins with registering reality or one ends up talking in circle."[409] The registering of reality is the point of departure of science as well as of natural theology. It is important for us to underscore Jaki's argument that registering reality places one on a linear progress and not a cyclic return. According to Jaki this is what Christianity has done to science for it to achieve the heights in which it has reached.

Care should therefore be taken that we do not mix-up what we think and what we are capable of knowing with the question of existence. What we know is limited by the conditions of our lives and our formation, we should therefore be cautious in passing from what we know to what should be. There are many ontological truths that lie beyond the capacity of the human mind that are not just mere thoughts. Besides, our knowledge is expressed in the language that does not completely express ontological realities or even carry the same meaning in some cases. The same language used to express scientific findings does not carry the same meaning theologically speaking; for example, the term "creation" in science is not the same

---

408 M. HELLER, *Creative Tension,* Templeton Foundation Press, West Conshohocken (PA) 2003, 146.

409 S. L. Jaki, *A late Awakening and other Essays,* 223.

as creation in theological terms, unless reconciled and intended to mean the same thing. This is because the theological order is not the same with the scientific experimental order. Jaki argues the verb "is", for example, expressed in the Bible as the name of God, carries a significance that goes beyond all ages in the philosophical and theological understanding.[410] We are therefore, invited, using our reason to give appropriate meanings to words we use at the various levels of abstractions.

The mystery of creation lies on the fact that science can explain all physical realities, but cannot explain why they exist and why there is a possibility of law in nature. The question of cause or origin will always occupy man not only on the physical level, but more pertinently on the metaphysical level. There is no time that we can possibly say that we have adequate knowledge of all that is. This means that we have no liberty to make definitive conclusions in the scientific realm. Reason should dread such conclusions.

The human mind that occupies itself with scientific endeavours is bound to encounter realities beyond itself. These realities cause in his thinking some "conflicts" because it is not produced by it. The human mind is challenged when it receives something from outside itself that it believes it is not produced by it. This may be a revelation attributed by the mind itself as an act of God. This is not intended to be a production of the mind, but an acknowledgment of the mind of realities that are found beyond itself. In the scientific field, this conflict is even more pronounced when science discovers that nature can only be fully understood by providing answers to metaphysical questions raised by it and which revelation seems to answer. The fact that the human mind is capable of arriving at this revelation is itself a manifestation of something greater than what we have at hand.

In dealing with science and religion, there is need to carefully use terms, because words have different meanings that makes the scientific language different from the religious language. We are limited by human language to express fully divine concepts and, as such, we are bound to intersect with both scientific and philosophical languages and fields. We have to admit that human language cannot fully express divine truths, and yet this limit

410 *Exodus* 3:14 (I am who I am).

must not prevent religion from using terms it may deem necessary to be religious. Words carry the meaning we assign them and, therefore, terms cannot be imprisoned to mean one thing by any one field of study, the context in which it is used must always be defined in order to avoid misconception. That is why one of the duties of philosophy is definition of terms. But also, scientists have to have a limit to the use of certain words, especially when their meanings carry with them concepts that go beyond its competence. Jaki argues: "If a scientist makes a valuational statement, he makes them as a philosopher and then he is obligated to justify his philosophy. This is a task which most scientists shy away from, or if they face up to it, they almost invariably make a mess of it."[411] He rightly makes a mess of it because he has nothing within his reach as a tool to defend himself outside the field of his competence. Rational freedom is found everywhere and yet intellectual boundaries are also founded everywhere.

3.12 Concluding Remarks

In this chapter we have discussed the rigorous use of reason as a requirement for the rationality of science and Christian faith. We have seen that reason should help the scientist move from facts into truth that lies outside the realm of science. We have also seen that this naturally occurs because of the ability given to man as a rational being. We have seen that the rigorous use of reason requires a sound philosophy. We quote again Jaki's concern that "the fundamental issue facing our scientific age is not about scientific know-how, but about the proper use of scientific knowledge and tools."[412] The problematic questions raised by science cannot therefore be responded to by science itself. We also put in evidence that the rationality of science stems directly from natural theology that has occupied the ancient man from the early philosophers to the modern time. Science and faith are reasonable because the universe rationally leads beyond the material realities. This, we have seen, is related to the fundamental question of existence and purpose. We also put into consideration that fact that ethical considerations

---

411 S. L. Jaki, *The Gist of Catholicism and other Essays,* 43.
412 S. L. Jaki, *Science and Creation*, Real View Books, New Hope (KY) 2016, p. xiv.

cannot be ignored in the propagation of science, because human free will is a given. We saw that scientist therefore cannot but must be responsible for their experiments, even when they are free to carry them out. We looked at the science and faith that it is a requirement for the progress of science. Scientist should depart from some form of belief both in themselves, in the tools that they use and in the fact that the cosmos will remain stable for their operations. Accusations that religion is based on unproven dogma of faith would not arise if science and religion take their formal objects seriously. We also put in evidence that some scientific discoveries were only possible because of the strong faith of the scientists who never gave up on the search of the truth. We also went ahead to discuss the limits of science that lead to faith and into metaphysical questions. We noted that science can only incompetently give answers to the serious questions their discoveries raise. We discussed some misconceptions about God and faith that arise and lead to an apparent misunderstanding of scientific objectives. We concluded the chapter by discussing faith and reason, emphasizing its use in relation to the advancement of science. We saw how the Catholic Church, accused forever of having blocked in history the progress of science, has always struggled through the magisterium to give a guide to scientific efforts without impeding it. We discussed that fact that it is the employment of sound use of faith and reason that will produce both a scientific and ethical giant. We put in evidence the fact that not all that is technologically possible is ethically practicable. The great question of our time will remain whether we shall be able to overcome, subjectivism, indifferentism, secularisation and the lack of the employment of rigorous reason in the discernment of scientific advancements in our modern times.

General Conclusion

Jaki has ably demonstrated the rationality of both science and the Christian faith by tracing the history of science and the relationship that should exists between science and religion. We have seen the vast approaches he has used from physical science to theology and philosophy. All these realities found in one man already shows the unity of rationality and the integration of knowledge. The conceptual separation can only be possible when the scope and limits are clearly set, bearing in mind that there is a possibility of encroaching them both intentionally and unintentionally.

When science reaches its limits, and where its methods cannot function, it should not stop asking questions, because this would go against the spirit of rationality. This is also applicable to conflict between science and religion in the sense that in the presence of apparent conflicts, and contradictions created by tension of strange opposition, theologians and scientists should not stop asking questions and seeking plausible resolutions to them.

It is only by constantly seeking answers to problems that religion and science will enrich each other. Care should be taken that no compromise is reached by creating false harmony, what Jaki refers to as "concordism" that brings more confusions than the harmony it intends to foster. This would be tantamount into a "sin" against reason.

It is important to guard against "totalitarian" tendency from both the side of science and religion. By totalitarian mentality I mean the desire of any of the two to deconstruct the other or to claim absolute authority of the definition of truth. The appeal should rather be to reasoned discourse and not to authority and wielding of power. This will make dialogue meaningful and with mutual benefit. The authority to command should come from an authentic interpretation of facts in relation to the discoveries of science and the doctrine of the faith supported by a strong philosophy that seeks the truth. Specialists in these fields must come to consensus or remain authorities only in their individual, relative fields. This means that one who has no qualification in both the scientific and the philosophical or theological fields should accept to hold a particular (partial) view, without pretending to propose it as universal and objective for others to follow.

This calls for an urgent development and openness to interdisciplinary or even multidisciplinary studies. The shortness of life, the vastness of the two fields and the human capacity may not allow many people to be experts in both science and theology and yet pertinent questions are supposed to be answered by those who can make good use of the wealth of both. It is evident that science and religion can be reconciled in the individual by proper use of reason and expertise leading to an integral knowledge of the given reality. This also confirms the fact that the two are neither exclusive nor opposed to each other, but can be complimentary in the sense that I have discussed in the second chapter of the dissertation.[413]

The evidence of the non-contradiction between science and religion is the fact that the two operate on the same nature given by God. The religious man knows God by reasoning systematically from the evidences provided by creation or better from nature. In the same manner, science operates, interprets and experiments on nature given to it by the Creator. Parting all from nature, notwithstanding revelation in the religious sense, science and religion are not in opposition, but are different fields of investigations that leads to different bodies of knowledge. It can be argued that religion has the duty to help science go beyond the physical world, and science has the duty to make man marvel about the greatness of nature and how it functions. Owen Gingerich, a professor of astronomy and the history of science at Harvard University once remarked: "I like the way these scientific explanations make sense. As a contemplative human being I am impressed by the creative achievements of humankind, [...] perhaps the purpose of the universe is simply for contemplative observer to exist in awe of a super intelligent designer; perhaps it is something more."[414] Experimental and mathematical science, honestly taken, points to some greatness that provokes in us the acknowledgement of a mystery so great. I would like to point out that the contemplative observer can be that one who is driven

413 Cf. H. H. OLIVER, "The Complementarity of Science and Religion", in J. D. PROCTOR (ed.), *Science and Religion in the Human Experience*, Oxford University Press, Oxford 2005, 143–159.

414 O. GINGERICH, "Is there design and purpose in the universe?", in J. F. HAUGHT (ed.), *Science and Religion in Search of Cosmic Purpose,* George Town University Press, Washington D.C. 2000, 131.

by awe and can be non-other than the philosopher who seeks to know for the sake of knowing.

We have also seen that methodological problems are major sources of conflicts between science and religion. Since the two have different ways of looking at the same thing, one investigating into its parts and another in harmonising and transcending it, care should be taken that these two ways are not confused or even used arbitrarily. It is, therefore, illicit to interpret theologically data from the laboratory out of its context and in the same way subjecting theological contents to laboratory tests. The point is the harmony between the two should be reached by developing a new method of studies. This new method, that is, listening to both science and religion, subjecting it to rigorous scrutiny and emerging out with a consensus, is one good way of resolving the problems between science and religion, that we have seen can only be done justice by the philosopher. We are aware of the dictatorship of the relativistic world but would like to insists that the truth be presented in and out of season.

Jaki has shown us the tendency of science to seek to eliminate religion or contradict its basic tenets in what he calls the "God of the fuse" mentality, what has commonly been referred to as the God of the gaps. I can assert that science must have as its intention the common good, or even specifically stated the good of man, since it deals with nature that the scientist has not brought into existence, since every part of this nature is link and related to each other and tempering with an element can jeopardise its totality. Explanations for apparent disconnections should not fall into relativistic approach to the question of knowledge but appeal to adequate methods that can answer unresolved issues.

There should be reasonable reasons for which a scientific experimentation or investigation is done. What is the rationale behind the experimentation? What ethical questions does the experimentation raise? What are the values and disvalues of the investigation in question? I strongly hold the view that scientists cannot afford to be indifferent to their investigations. It is vain glory, for example, to desire to construct an office in one of the planets or to continue the search for extra-terrestrial intelligent beings who are similar to us. Scientists cannot continue to experiment on stem cells knowing very well that, to do so, they must kill an unborn baby or pretend to be indifferent when an experiment is to use another person as a means.

This is the wisdom that Pope John Paul II calls for in his Encyclical *Fides et Ratio:*

> In expressing my admiration and in offering encouragement to these brave pioneers of scientific research to whom humanity owes so much of its current development, I would urge them to continue their efforts without ever abandoning the *sapiential* horizon within which scientific and technological achievements are wedded to the philosophical and ethical values which are the distinctive and indelible mark of the human person.[415]

Man by his very nature is bound to discern the necessities and the consequences of his actions. The Pope even points out that these elements are wedded to each other, meaning that they are indispensable for any meaningful scientific endeavour, and for as long as man does it in a reasonable way. There must be the ethical and philosophical face of the reality under investigation. The *sapiential* horizon calls for a keen view of how multiplicity is intertwined and interwoven in each other to form a complete whole, good and unique. "Our minds should work in the light of reason as well as in the light of faith."[416] We have seen in Jaki's arguments that even the scientist has to have some form of faith, which should not be taken as the same faith of a Christian, in order to be successful. He has to in some way believe in his ability to carry out experiments, the tools that he uses, the stability of nature, the constancy of his experiments and that the laws of nature will remain the same today and tomorrow. These are realities for which there are no scientific proves and yet they are reasonable.

One of the greatest achievements of Jaki was the ability to demonstrate the reality of the universe and giving it a very particular definition as the totality of constantly interacting things in motion. Even if the term universe is an abstraction from all that is, it has a firm base on concrete realities whose totality and mode of operation produce it. I think everything in the universe is important for its life and, if there is anything done with it, it should be to put into good use the values it carries and not to destroy even a single of its elements. It is this value and meaning that leads the scientist

---

415 John Paul II, *Fides et Ratio*, n. 106.
416 G. M. Verschuuren, *God and Evolution; Science Meets Faith,* Pauline Books and Media, Boston (MA) 2012, 5.

to the acknowledgement of the role of religion and the faith that it comes with in the life of all humankind.

At this point, we should ask ourselves a fundamental question: how should religion act in the midst of scientific progress? It is therefore, plausible to assert that religion can neither afford to ignore nor lag behind the scientific progress. It is not even enough to make calculated statements about science, cosmology and technological advancement. There is a dire need of training scientists who can ably unite the theological and scientific field so that a meaningful dialogue and integration is promoted. This is the only antidote for scientism and religious bigotry.

I also wish to point out Jaki's argument that rationality is a moral choice and consequently to suppress reason, ignoring its use or acting irrationally is the same as acting immorally. It is poignant to observe that our world today is not enthusiastic about putting rationality to use but ready to absorb all irrationality without minding or filtering it. The choice of indifference to both the use of reason and life endeavours are great dangers for our society today. The first step before any scientific investigation should be to argue convincingly, demonstrating the pros and cons of the endeavour, defending rationally one's stand and justifying why the particular scientific initiative is worth undertaking and should be allowed to go ahead. This is a requirement for the rationality of science and faith. I think it is evident that the challenge of rationality is to act according to the dictate of right reason. I believe that no scientific experimentation or investigation should be engaged into by chance or mere curiosity without a global view of its consequences, be it social, ethical or bioethical.

Scientists should also have the humility to admit that the question of origin of all that is cannot be answered ably by science alone; it begs for another level and type of knowledge which human reason is capable of. There is, therefore, need to obey reason here, because it is capable of metaphysical knowledge. Science can remain one of the sources of this knowledge. The guiding questions will always remain: Why is the universe so orderly? Why is scientific investigation possible? Why should there be something and not nothing? Why does our reason go beyond itself? Why is reason not satisfied with the physical and sense knowledge only? Why is probability possible? These are questions that do not need a one sentence or one-word answer but are pointers to realities beyond our physical world and capacities. The awareness of our limits is one great step towards an intellectual humility that lacks no rigour.

I can say without doubt that we cannot continue to say that the universe is here because we are here, and that we can make of it whatever we want. This universe has an author whose operation in it cannot be ignored. Reason feels fulfilled when it acknowledges this cause beyond the human being, beyond chance and chaos. This is one of the reasons why Jaki believed that science could only be born after a good understanding of the difference between creation and creator. Our reason tells us that the universe is authored by an intelligent personal God. Pope John Paul II in *Fides et Ratio* writes: "Scientists are well aware that the search for truth, even when it concerns a finite reality of the world or of man, is never-ending but always points beyond to something higher than the immediate object of study, to the questions which give access to mystery."[417] The truth of nature points to the truth *par excellence* even in its most simple and ordinary form. There is that continuity that flows from nature to the creator and ignoring it is tantamount to intellectual dishonesty.

The question of evolution is one of the areas in which many philosophers and scholars use to attack religion and faith: one can think of men like Dawkins, whom we have referred to time and again, because of the nature of his publications.[418] It is true that the attacks on God create curiosity in every man, and thus the attraction to buy books that talk about God. Biology alone cannot explain the origin of both the material and immaterial things in the universe. "To evolve literally means 'to unroll a scroll,' that is, to read a book. The imagery of nature as a book has its roots in Christianity and has been held dear by many scientists."[419] The theory of evolution implies that great questions must be assigned to philosophy. These questions, being assigned to philosophy, pushes them beyond the

---

417 J. Paul II, *Fides et Ratio,* n. 106.

418 Cf., R. Dawkins, *The God Delusion*, Bantam Press, 2006.

419 Benedict XVI, "Address of His Holiness Benedict XVI to the Members of the Pontifical Academy of Sciences" (31, October 2008), in W. Arber – N. Cabibbo – M. S. Sorondo (eds.), *Scientific Insights into the Evolution of the Universe and of Life,* Pontifical Academy of Science, Vatican City 2009, xxxiv.

scope of the natural sciences. Therefore, if anyone is to argue against religion using evolution, such a person must be ready to engage in serious philosophical discourse, that does not only demand a single philosophical view, but an entire philosophical tradition. Evolution, used as an explanation is of the origin of existence, parts from true science into scientific ideology as argued by Jaki.

I am of the view that scientific discussions go beyond both mere religious piousness and pure scientific method to an honest use of reason that manifests itself both in nature and in the thinking subject. Both man and the universe are reasonable. Philosophy is required as an entity mediating between faith and the natural sciences. "Philosophy is sought in order to formulate the limits of the scientific methods and their scope, in order to reveal boundary crossing, in order to open any narrowed concept of reason."[420] The role of philosophy is enormous in harmonizing the relationship between science and religion. It becomes the pivotal point of unity and the meeting point in which the conflicts between science and religion are resolved, dialogue promoted and intellectual integration achieved. It is through philosophy, guided by proper reason, that relevant metaphysical questions can be asked and reasonably digested to explain the origin of the universe.

It is also important to look at the role of authority in the relation between science and religion. I must take a distance from the affirmation that the Catholic Church, at one point in history, took a totalitarian position in confrontation with science, especially when the Catholic Church had to deal with the Galileo Galilei's case and the evolution theory.[421] Jaki has ably

---

420 C. Schönborn, "Pope Benedict XVI, On Creation and Evolution", in W. Arber – N. Cabibbo – M. S. Sorondo (eds.), *Scientific Insights into the Evolution of the Universe and of Life,* Pontifical Academy of Science, Vatican City 2009, 19.

421 We should of course admit that it has never been easy for the Church to discern the fast-changing achievements of sciences. Creation of various groups, such as the Vatican Observatory and the Academy of the Sciences, to follow scientific and cosmological progresses, was a great effort on the side of the Church. The intention of the Church has always been to guide scientific research to a philosophy that promotes the goodness of man and the knowledge of God. Man's wisdom should bond him more with God than separate him.

demonstrated the complexity of the Galileo question and how both sides could have avoided the clash. The authority of the Catholic Church has always been in defence of true interpretation of scientific data, in relation to scriptures and her doctrine, of which she is a custodian. Pope Pius XII has long noted in relation to evolution that the magisterium does not forbid discussions and research on evolution of the body from pre-existent and living matter. He however warned that this be done with moderation and respect for authentic interpretation of scriptures.[422] It is evident that there is an openness to scientific advancement but the Catholic Church insists that science should take into consideration that there are other truths that go beyond it. The question of the human soul as being the same at conception as created directly by God cannot be subject to evolution theory that has no tool first of all to affirm or deny the existence of the soul. I am convinced that the guiding line for both science and religion is always the truth that should prevail. I think that this truth is not manufactured or created by anyone, but presented to us in nature and revealed by God. Research of any kind must have as its base the honest and selfless search for the truth.

Pope Benedict XVI, in his address to the members of the Pontifical Academy of Sciences, on 31st of October 2008, says:

> To state that the foundation of the cosmos and its development is the provident wisdom of the Creator is not to say that creation has only to do with the beginning of the history of the world and of life. It implies, rather, that the creator founds these developments and supports them, underpins them and sustains them continuously.

This is an appeal against any form of reductionism; taking creation only from an historical view and ignoring the wisdom that such implies, would only be looking at the effect of what was just in existence. In dealing with the rationality of faith and science, we do not only need intelligence but also the wisdom to discern and know their limitations and differences.

---

422 Pius XII, Encyclical Letter *Humani Generis* (12 August 1950), n. 36.

Appendix

Password for Catholic Christians

After my defence of the thesis; *The Rationality of Science and the Rationality of faith according to Stanley Jaki*, I visited a wine factory. A friend of mine who organized the visit kept introducing me to people and announcing that I had freshly obtained a PhD in philosophy with the above thesis. The more I tell him not to introduce me initially as a priest and doctor and reserve it until we are about to leave, the more he did it. My intention is always to meet people in their natural setting without making them adjust to status and titles. However, this became an opportunity for me to respond to the questions about science and faith that linger in the minds of many Christians. Many of these questions are really misconceptions and lack of a rigorous use of reason. I have decided to dedicate this short section in offering a guide to Christians who are confused when facing the advocates of experimental science who see this as the saviour of the world.

The first man I met that morning had greeted us unenthusiastically in his wine factory. We had popped in to see where my friend formerly worked and he was expecting some more organised groups. After some general information and a cold welcome, my friend told him about my thesis. The mood instantly changed and a discussion ensued that lasted over 40 minutes, an equivalent of a lecture period! The first reaction of our host was, "can religion be rational?" He then quickly went into the justification of his position by asserting that religion has brought suffering and exploitation in the world. It is a way of oppressing people and keeping them docile under the heavy hands of their oppressors, etc. I noticed immediately Karl Marx's ideology that "religion is the opium of society" emerging from his arguments. Most of the time people who oppose religion have immediate responses that forces any ordinary Christian to resort to saying, "well that is according to you." That kind of relapse to helplessness is a shame on the part of one who should justify the reason of his belief. We must learn to ask the right questions so that we do not guard against bullets that are not fired at us and be ready to defend our position. I asked our host a very simple question, "Which religion are we talking about?" He was surprised because he was talking about religions in general. Then he retorted, "even Christianity does the same." I then told him, let us talk about Christianity then. He also confessed that he is a Christian but not a practising one.

Then I asked him what are the fundamental doctrine of the Christian faith? What are the tenets? What is the viewpoint of the social teachings of the church on all the problems raised? Why does the Catholic Church have a justice and peace commission? My good friend did not have answer. Then I told him that maybe those questions are too farfetched and I would ask him a question stemming from his own statement; "Which of the Christian doctrine teaches exploitation, and war or teaches Christians to be stupid and docile in the face of injustices?" He was again at a loss for words and said even Christians at one point waged war in the names of crusade. My next question marked the end of the conversation. Was it Christianity that waged war in the name of crusade or Christians? His answer was definitely instant "Christians!" "Well said" I replied? "then, when Christians do not follow what Christ has left for them as a model to follow, it is not Christianity that fails but Christians. When Christians fail to meet the rationality required of their faith, or do not discern well the requirement of their faith, it is not the failure of Christianity but the failure of Christians." We must learn to distinguish practice from content. It is what I refer to as 'a user problem.' You see, I avoided falling into the discussion of the crusade that has its own history, justification and time. If we must judge history, we must do it within the context of its time.

To answer his question, I told him that Christianity has clearly defined doctrine and a rational content of what should be believed and summarised in short sentences called 'The Creed.' The belief in God as creator, saviour and sanctifier is clearly defined so that the Christian faith is based on strong reason as opposed to fideism. Our discussion ended by the owner of the wine factory saying, "It is interesting to talk about these things. We can organise another time and discuss them over a glass of wine." Before you begin to defend yourself against any attack, define the problem and make sure you are talking about the same God, less you end up defending a God you do not believe in.

Meeting with an Atheist

On another occasion I was in restaurant and a woman was talking loudly on phone in English in a place where she thought no one would follow her conversation because they spoke Italian. She was advising a son not to accept his fiancée's proposal of marriage. Her argument was that matrimony is a creation of the church and the society and that she never got married herself and she is living happily with her partner whom she referred to as her 'companion.' When she realised I was following her conversation, she turned to me and asked whether I spoke English, to which I responded in the affirmative. She then felt the obligation to justify herself and declared that she was an atheist and would not accept that her son should get married because of the reasons she had stated. This then led to an interesting and complex discussion. I asked how old her son was and she proudly said, "my boy is thirty now!" I retorted, "that is great! He is a big boy now!" I asked her another question, "Where does your boy live?" "He is in London," she replied. "Is he employed?" "Yes!" she responded. If he is employed, that means he works with people and lives with them. Am I right? She answered, "definitely!" Then it was time to ask the right question, "Did you say matrimony is a creation of society, that same society in which your son lives and works?" She became defensive and asked, "What does that have to do with the matrimony of my son when I am an atheist?" I did not want to let the discussion deteriorate to sentimentality and I retorted, "Your son lives in a society that acknowledges the union of a man and woman in matrimony and yet your son is free to get married or not because marriage is a matter of free consent." Then I added that I would ask her one more question about her son and we shall talk about her being an atheist. She quickly accepted and I asked, "How many days of the life of your son will you live for him or more pertinently how many days of the 30 years of his life did you live for him?" At this point there was no answer and I told her, "I think you should allow your son to live his life in full freedom since the consequences of his choices shall bear upon him and his life." She retorted, "We can now talk about atheism which is my choice. I am an atheist and my parents were atheist and I raised my son as an atheist."

When you are confronted with such affirmations, you need to keep calm and be as friendly as possible. You do not need to begin to impose your creed on anyone. Catholic Christians must learn the art of rigorous thinking so that the content of their faith is knowledge not empty belief, "Faith without reason is like a bird with one wing," as St. John Paul II wrote in Fides et Ratio explaining that faith and reason are like two wings that help the human person to fly to great heights of the mysteries revealed in Jesus.

I expressed my great ignorance to the old woman and asked her to explain to me what she meant when she talked of atheism and requested that she could talk about atheism without making a reference to God. She told me that she could not do this. I told her that it is possible since atheism is a negation of theism, and she can talk about it. She then said that she does not believe in the existence of God because there is no God. Most atheists would articulate this and that is acceptable. You do not need to make him believe in God but you need to establish that you are talking about the same God you know. I then asked her to define the God that she does not believe in and the characteristics of that God. She was quick to say that she does not believe in anything called God. Then I told her that it is ok that she does not believe in anything called God but she denies something called God. I then asked her to talk about that something whose denial makes her an atheist. She was unable to respond "I am having headache with your questions, let us stop it there." Did she have the justification of what she is proclaiming? Maybe not, but she is not different from a Christian who has no rational defence of his faith.

I did not have to defend my God because there was no attack on the God of history revealed in time in Jesus Christ, who is a person who relates to creation and continues to operate in it. An intelligent atheist would build on the problem of evil and concludes that a good-great-God should not permit evil if he exists. This will not be the Christian God because the Christian God did not come in Christ to remove evil, which would ultimately be the removal of human freedom. He came to teach us how to overcome evil and to accompany us in the face of evil that makes us creatures.

Meeting a Believer in Science

Another time a 16 year old child called Matthew follow me to the sacristy and pronounced "Father, I do not believe in God, I believe in science that solves all the problems of the world and not God." I admired his boldness and asked him to talk about science. He was able to talk about all the capabilities of physical science, technology and discoveries. Then I asked him to talk about the God he does not believe him. He could only affirm that God does not exist but could not speak about that which does not exist at least in the form in which those who hold his existence do. I advised him to learn more about the what he does not believe in so that he could argue more successfully against it.

We had to go back to the foundation of his presumed content with systematic interrogations. I asked the young man of the origin of all the things that science deals with. Where do these things come from? How is it that they are there? How did they come to be? These are definitely causal questions that science covers too, even when science deals with cause and effects for which they accumulate data and draw conclusions. The question is that they deal with something whose cause lies outside the scientist's fabrication. At this the young boy was at a loss for words because he was too young to comprehend the ultimate cause of all things. I understood later that he was indoctrinated by his Dad about the faith in science.

Are you a Christian? Learn to ask the right question before beginning to defend yourself against anyone who opposes your faith. Before you begin an argument about your faith agree with your opponent that you will make reason your common denominator something you both must possess to begin the discussion in the first place. If reason is not the common denominator, I would advise that you talk about the weather and restaurants in which you can find good food and drinks. Science and faith demand intellectual rigour without which science becomes scientism and faith becomes fideism. Be on your guard then.

Bibliography

1. Works of Stanley Jaki

The Relevance of Physics, University of Chicago Press, Chicago (IL) 1966.

Brain, Mind and Computers, Herder and Herder, New York 1969.

The Paradox of Olbers' Paradox: A Case History of Scientific Thought, Herder and Herder, New York 1969.

The Milky Way: An Elusive Road for Science, David and Charles, New York 1972.

Planets and Planetarians: A History of Theories of the Origin of Planetary Systems, Scottish Academic Press, Edinburg 1978.

The Origin of Science and the Science of its Origin, Gateway, South Bend (IN) 1978.

Cosmos and Creator, Scottish Academic Press, Edinburgh 1980.

"The Case for Galileo Rehabilitation", in *Fidelity*, March, 1986, pp. 37–41.

Chesterton: A Seer of Science, University of Illinois Press, Chicago (IL) 1986.

Uneasy Genius: The Life and the Work of Pierre Duhem, M. Nijhoff, Dordrecht 1987.

The Purpose of it all, Scottish Academic Press, Edinburgh 1990.

Scientist and Catholic: Pierre Duhem, Christendom Press, Front Royal (VA) 1991.

Patterns or Principles and Other Essays, The Intercollegiate Studies Institute, Bryn Mawr (PA) 1995.

God and the Cosmologists, Real View Books, Fraser (MI) 1998.

Bible and Science, Christendom Press, Front Royal (VA) 1999.

Means to Message: A Treatise on Truth, William B. Eerdmans Publishing Company, Grand Rapids (MI) 1999.

The Limits of a Limitless Science, ISI books, Wilmington (DE) 2000.

The Saviour of Science, William B. Eerdmans Publishing Company, Grand Rapids (MI) 2000.

The Gist of Catholicism and Other Essays, Real View Books, Pinckney (MI) 2001.

A Mind's Matter; An Intellectual Autobiography, Grand Rapids (MI) 2002.

Questions on Science and Religion, Real View Books, Pinckney (MI) 2004.

The Drama of Quantities, Real View Books, Port Huron (MI) 2005.

The Road of Science and the Ways to God, Real View Books, Port Huron (MI) 2005.

"The Case for Galileo's Rehabilitation", in LARREY Philip, *Science and Religion: An Anthology, Selection of Texts from Aquinas, Bacon, Galileo, Darwin and John Paul II*, ARCANE (ed.), Rome, 2006, pp. 254–263.

A Late Awakening and Other Essays, Real View Books, Port Huron (MI) 2006.

The Impassable Divide, Real View Books, New Hope (KY) 2008.

Lectures in the Vatican Gardens, Real View Books, New Hope (KY) 2009.

The Mirage of Conflict Between Science and Religion, Real View Books, Fraser (MI) 2009.

Science and Creation, Real View Books, New Hope (KY) 2016.

2. Classics

PLATO, *Great Dialogues of Plato*, translated by ROUSE W. H. Denham, edited by WARMINGTON H. Eric – ROUSE G. Philip, New American Library, New York, 1956.

ID., *The Complete Works*, translated by JOWETT Benjamin, Oxford University Press, Oxford 2011.

ARISTOTLE, *The Complete Works*, edited by BARNES Jonathan, Princeton University Press, Princeton (NJ) 1991.

ID., *Aristotle Physics; Book VIII*, translated by Daniel Graham, edited by ACKRILL J. – LINDSAY Judson, Clarendon Press, Oxford, New York 1999.

THOMAS AQUINAS, *Summa Theologiae*, translated by BAUMGARTH P. William and REGAN J. Richard, University of Scranton Press, London 1994.

GALILEO Galilei, *Selected Writings*, translated by SHEA R. William – DAVIE Mark, Oxford University Press, Oxford, New York, 2012.

3. Magisterium

PIUS XII, Encyclical Letter *Humani Generis* (12 August 1950).

JOHN PAUL II, Encyclical Letter *Fides et Ratio* (14 September 1998).

4. Studies

AGNOLI Franceso, *Scienzati, Dunque Credenti; Come la Bibbia e la Chiesa hanno creato la Scienza sperimentale,* Cantagalli S.r.l., Siena, 2020.

ARBER Werner – CABIBBO Nicola – SÁNCHEZ S. Marcelo (eds.), *Scientific Insights into the Evolution of the Universe and of Life,* Pontifical Academy of Science, Vatican City 2009.

ASHLEY M. Benedict, *The Way Towards Wisdom,* University of Notre Dame Press, Notre Dame (IN) 2006, EPUB e-book.

BARBOUR G. Ian, *Religion and Science: Historical and Contemporary Issues*, Harper & Row, San Francisco (CA) 1997.

ID., "Ways of Relating Science and Theology," in R. J. RUSSELL – W.R. STOEGER – G.V. COYNE (eds.), *Physics, Philosophy and Theology: A Common Quest for Understanding*, Vatican 1988, pp. 21–48.

BARR M. Stephen, *The Believing Scientist,* William B. Eerdmans Publishing Company, Grand Rapids (MI) 2016.

ID., *Modern Physics and Ancient Faith,* University of Notre Dame Press, Notre Dame (IN) 2003.

BOUDRY Maarten – PIGLIUCCI Massimo (eds.), *The Challenges of Scientism*, University of Chicago Press, Chicago (IL) 2018.

BULTMANN Rudolf, *Jesus Christ and Mythology,* Charles Scribners Sons, New York 1958.

BUTTERFIELD Herbert, *The Origin of Modern Science*, Free Press, New York 1957.

CONGIUNTI Lorella, "La collocazione epistemologica del "caso": all'incrocio dei saperi" in *Urbaniana University Journal*, 1 (2015), pp. 109–140.

CONNELL J. Richard, *Nature's Causes,* Peter Lang Publishing, New York 1995.

ID., *The Empirical Intelligence – The Human Empirical Mode; Philosophy as Originating in Experience,* Lewiston, New York 1988.

ID., *Substance and Modern Science,* The Center for Thomistic Studies, Houston (TX) 1988.

ID., *From Observables to Unobservable in Science and Philosophy,* University Press of America, New York 2000.

CRAIG William Lane, *God and Abstract Objects. The Coherence of Theism: Aseity*, Springer, Berlin 2017.

CROMBIE ALISTAIR Cameroon, *Robert Grosseteste and the Origins of Experimental Science*, Clarendon Press, Oxford 1953.

ID., *Styles of Scientific Thinking in the European Tradition*, 3 Vol., Duckworth, London 1994.

DAWKINS Richard, *The God Delusion*, Bantam Press, London 2006.

DAWSON Christopher, *Religion and the Rise of Western Culture,* Sheed and Ward, London 1950.

DEAN Cornelia, *Making Sense of Science. Separating Substance from Spin*, Harvard University Press, Cambridge (MA) 2017.

FORMICA Giambattista, "L'assiomatizzazione della matematica e le ricadute dei teoremi di Gödel", in P. PECERE (a cura di), *Il libro della natura*, vol. 2, Carocci, Roma 2015, 229 264.

GILSON Etienne, *The Unity of Philosophical Experience,* Charles Scribner's Sons, New York 1937.

GIOSTRA Alessandro, *Stanley Jaki: Science and Faith in a Realist Perspective,* IF Press, Rome 2019.

GRANT Edward, *The Foundation of Modern Sciences in the Middle Ages: Their Religious, Institutional and Intellectual Contexts*, Cambridge University Press, Cambridge 1996.

HAFFNER Paul, *Creation and Scientific Creativity: A Study in the Thought of S. L. Jaki,* Gracewing Publishing, Leominster 2009.

HAUGHT F. John, *Science and Religion in Search of Cosmic Purpose,* Georgetown University Press, Washington D.C. 2000.

HELLER Michael, *Creative Tension,* Templeton Foundation Press, West Conshohocken (PA) 2003.

HOOYKAAS Reijer, *Religion and the Rise of Modern Science,* Scottish Academic Press, Edinburgh 1972.

HUFF E. Toby, *The Rise of Early Modern Science: Islam, China and the West*, Cambridge University Press, Cambridge 1995.

KAISER Barina Christopher, *Creation and the History of Science*, Marshall Pickering, London 1991.

LARSON Edward J. – RUSE, Michael, *On Faith and Science*, Yale University Press, New Haven (CT) 2017.

LINDBERG C. David – NUMBERS L. Ronald (eds.), *God and Nature: Historical Essay in the Encounter Between Christianity and Science,* University of California Press, Berkeley (CA) 1986.

ID., *The Beginning of Western Science. The European Scientific Tradition in Philosophical and Institutional Context BC-AD 1450,* University of Chicago Press, Chicago (IL) 1992.

ID., *The Cambridge History of Science*, 7 Vol., Cambridge University Press, Cambridge 2002–2018.

LINDBERG C. David, (ed.), *Science in the Middle Ages,* University of Chicago Press, Chicago (IL) 1978.

LIVANIOS Vassilis, *Science in Metaphysics: Exploring the Metaphysics of Properties and Laws*, Springer, Berlin 2017.

MENÉNDEZ Guerra – PASCUAL Raphael – COLOMBO Antonio (eds.), *Science and Faith in Stanley Jaki. Proceedings of the Summer Course*, If Press, Rome 2017.

MONOD Jacques, *Chance and Necessity,* translated by WAINHOUSE Austyn, Vintage Books, New York 1972.

NEWMAN John Henry, *The Grammar of Assent,* Assumption press, Dayton (OH) 2013.

O'CONNOR Daniel – OAKLEY Francis, *Creation: The Impact of an Idea*, C. Scribner's Sons, New York 1969.

OZIOKO Uchenna Johnson, *The Rationality of the Christian Faith in Richard Swinburne,* Urbaniana University Press, Vaticano, Rome 2019.

PROCTOR D. James, *Science, Religion and the Human Experience*, Oxford University Press, Oxford 2005.

RELJA Hrvoje, *Realismo di S. L. Jaki. Dalla convinzione religiosa tramite il realismo moderato e la creatività scientifica fino al realismo metodico,* PhD, Dissertazione, Ateneo Pontificio Regina Apostolorum, Roma 2008.

SAMBURSKY Samuel, *The Physical World of the Greeks*, Princeton Press, Princeton (NJ) 2014.

SORABJI Richard (ed.), *Philophonus and the Rejection of Aristotelian Science,* Duckworth, London 1987.

TRASANCOS Stacy, *Particles of Faith: A Catholic Guide to Navigating Science,* Ave Maria Press, Notre Dame (IN) 2016.

ID., *Science was Born of Christianity: The Teachings of Fr. Stanley L. Jaki*, The Habitation of Chimham Publishing Company, Titusville (FL) 2014.

VERSCHUUREN M. Gerard, *God and Evolution? Science Meets Faith*, Pauline Publishers, Boston (MA) 2012.

WILSON O. Edward, *Sociobiology: The New Synthesis*, Harvard University Press, Cambridge1975.

5. Internet Source

BARR M. Stephen, https://www.firstthings.com/article/2009/06/science-seeking-understanding.

MANDER William, "Pantheism", *The Stanford Encyclopedia of Philosophy* (Spring 2020 Edition), Edward N. Zalta (ed.), URL = <https://plato.stanford.edu/archives/spr2020/entries/pantheism/>.

www.ingramcontent.com/pod-product-compliance
Ingram Content Group UK Ltd.
Pitfield, Milton Keynes, MK11 3LW, UK
UKHW041826031026
15587UKWH00002B/584

9 783631 876114